MAN
MASTERS NATURE
Twenty-five Centuries
of Science

MAN
MASTERS
NATURE

Twenty-five Centuries
of Science

Edited by Roy Porter

George Braziller

NEW YORK

Published in the United States in 1988
by George Braziller, Inc.

First published in England in 1987 by BBC Books, a division of
BBC Enterprises, 80 Wood Lane, London W12 0TT

George Braziller, Inc.
60 Madison Avenue
New York, New York 10010

Library of Congress Cataloging-in-Publication Data

Man masters nature.

   To accompany the BBC radio series Man masters
nature first broadcast from spring 1987.
   Bibliography: p.
   Includes index.
   1. Science—History.   2. Scientists—Biography.
I. Porter, Roy, 1946–
Q125.M3165   1988   509   87-32738
ISBN   0-8076-1193-X

Printed in the United States
First printing

CONTENTS

Man masters nature not by force but by understanding. That is why science has succeeded where magic failed: because it has looked for no spell to cast on nature.

Jacob Bronowski
Science and Human Values, 1956

INTRODUCTION

If science is the knowledge of, and power to control Nature, then all societies have possessed science. Without some grasp of the orderly cycle of the seasons and the climatic changes accompanying them, of the habits of animals and the properties of basic materials, of what makes people sick or strong, no tribe, even the most 'primitive', could ever have survived. Some societies, however, have transformed such hard-won experience into system, into theory, into bodies of truth, which satisfy the human need to understand as well as to act. Here, looking far back, we can point to Babylonian or Egyptian science, or, more recently, to Hindu, Islamic or Chinese science. Many of these traditions of understanding Nature – the Chinese one in particular – have been immensely impressive.

Yet there is something unique about Western science, that body of facts, theories and practices first systematised by the Greeks, revitalised by the Renaissance, revolutionised by the so-called 'New Science' of the sixteenth and seventeenth centuries, and advancing at an astonishing rate ever since. For good or ill, European or, latterly, Western civilization has come to master the globe over the last five centuries. And no small part in that conquest – a conquest of the environment, of other empires, and not least of minds – has come through the unique strength of Western science.

Science teaches us to be critical and objective, and so warns us of the dangers of thinking our own viewpoint is automatically the correct one. And philosophers have rightly insisted that we have no way of proving that our science is true, and is the only truth. So it would be presumptuous and short-sighted to dismiss everything else as ignorance and error, and to assume that our science possesses a monopoly of truth, still less of wisdom. All the same, all our experience daily tells us of the unmatched power of our science – it can put a man on the Moon, it can eradicate smallpox from the world, it can, indeed, eradicate *life* from the world. Western science has now become world science. Quite distinctive traditions of painting, poetry, religion, philosophy and so forth continue to thrive throughout the globe. But, inside the laboratory in Peking or Penang, Lima or Lagos, it is Western science which is restlessly at work.

What has made Western science so uniquely successful? That is a key question lying behind this book. The authors of the chapters that follow collectively address themselves to showing how and why our science never ossified, never became a dead-end (unlike, say, alchemy or much that passed for metaphysics), but rather continually renewed itself, extended its intellectual credibility, and met the needs of industry, technology and society. Those essays will speak for themselves. But a few broad comments first will help to set the scene. The keynotes are *unity* and *comprehensiveness*.

Science as we know it, is now, and long has been a bundle of different disciplines – physics, chemistry, botany, zoology and so forth – broadly the physical sciences, the life sciences, and the human sciences. Newer, narrower specialisms have recently proliferated, such as endocrinology and genetics. This division of labour is inevitable and efficient. But, above all, our science has been extraordinarily *unified* and *intergrated*. We owe the inspiration for this to the Greeks. When they put their minds to the mass of bitty facts and speculations they inherited from their predecessors, Greek thinkers saw it as one of their prime tasks to reduce all to order. The Ionian philosophers and then, above all, Plato and Aristotle believed that human reason could show that Nature was sublimely orderly; it obeyed its own laws, rather than being subject to the caprice of the gods; it was made up of a small number of basic materials (some thought 'elements', others 'atoms'), ultimately linking the universe (macrocosm) to man (microcosm); it was regular in its operations.

If Nature was rational, man could comprehend it, and so science was possible. Greek science set about bringing Nature to order. Geometers stressed the importance of form, number and harmony. Aristotle showed the fundamental unity of all things living (and much more besides); Ptolemy undertook a comprehensive mapping of the heavens and the globe, establishing a practical astronomy and geography united by the practice of measurement. The theories of Aristotle, Ptolemy and a small number of other giants of thought, made science a form of knowledge – systematic, rational, abstract – which could be stored, taught and transmitted, ultimately criticised, and so advanced.

It was thus one of the great strengths of the scientific tradition bequeathed by the Greeks that there were always clear principles to grapple with. Being so coherent, Greek science was passed down in a clearly-packaged form for the next 2000 years. Thus when in the seventeenth century, William Harvey puzzled out fundamental problems of how the body organs worked, not least how the heart and the blood were related, he found he could go back to Aristotle and bounce his ideas off him. For Harvey, Aristotle was neither a god-like authority, never to be challenged (that attitude stultifies science), nor was he an antiquated buffoon. Rather he was a mind that mattered, an intellect to engage. Even Harvey's contemporary, the inconoclastic Galileo, who made such fun of benighted Aristotelians, got more from Aristotle than he was prepared to admit. Science has succeeded through keeping tradition and innovation in creative tension. From the Greeks onwards, Western science has avoided the 'clique-iness' of occult sects, the esotericism of magic and mysticism. By being a public body of facts and theories, it has always been open to improvement.

This comprehensive quality, which hinges upon the idea of *Nature as a unity*, has paid dividends in another dimension. Science has progressed by specialisation. But such diversification has often eventually led to new and unexpected reunification, demonstrating a higher order of coherence and opening up fundamental new fields. For example, two distinct sorts of astronomy developed after Ptolemy. One practical and mathematical, excellent at predicting the positions of the planets but making little physical sense. The other formed a body of cosmology, which was largely qualitative and speculative. In an act of radical imagination and great daring, Kepler in the seventeenth century was able to reunite these two strands and develop a true physics of the heavens. Or, to take another example, after Newton, the various physical sciences made their separate advances: light, heat, magnetism, electricity, and so forth. But particularly from the nineteenth century, the great scientific advances came through demonstrating a higher level unity of these apparently distinct phenomena. Through the work of Watt and Black, and Carnot and Joule, mechanics and heat eventually became integrated in the thermodynamics formulated by Lord Kelvin.

Similarly, a succession of great minds from Davy and Faraday, to Einstein and Bohr, demonstrated that magnetism, electricity, energy, light and gravity were not just separate phenomena, but were all intimately interconnected. Geography, geology, palaeontology and embryology all ploughed their own individual furrows in the nineteenth century. Not the least of the achievements of Darwin was to perceive how all these specialisms could actually fertilise each other – indeed, that they made sense ultimately only within a unifying theory of evolution. The same also applies to the human sciences. At first glance, computers may seem merely a form of higher gadgetry, a piece of engineering wizardry. But the very idea of the computer, as it gelled in the mind of Alan Turing, arose from profound speculations upon the precise possibilities and limitations of the human mind, in the light of a fundamental dilemma about the status of mathematics. The pursuit of the particular, but its integration within the universal, is surely another key to the vitality of our science.

Because of this fascination with unity and wholeness, it comes as no surprise that so many of the master scientists have possessed a powerful aesthetic, mystical, or religious sense of the harmony of the cosmos, or of Divine Creation. Einstein distrusted Bohr's quantum theory because it threatened to reduce Nature to arbitrariness. 'God does not play dice with Nature', Einstein concluded. Kepler fundamentally saw it as his religious vocation to discover, and then to reveal to his fellows, that higher order of Nature which God had created and which He had waited for so long for man to comprehend. Newton, Priestley, Faraday and many others were guided towards their syntheses of matter and spirit, atoms and elements, forces and fields, by powerful – sometimes Faustian – philosophical and religious impulses, driving them to penetrate the secret ways of the Maker.

All this may seem a far cry from certain popular images of the scientist, regarded as a dry-as-dust, detached, dispassionate investigator, amassing facts, weighing theories. In a highly influential philosophical account of science, Sir Karl Popper has presented us with an image of the scientist proceeding by setting up theories and then trying to prove them all wrong ('falsifiability'). Moreover, in Britain we have been inclined to think in terms of 'two cultures', on the one hand, the humanities, fired by imagination and creativity, on the other, science, often seen as objective and neutral.

But these really amount to gross caricatures of the scientist. The history of science will illuminate this and, in turn, will make no sense unless we grasp the very special nature of the working, thinking, scientist as he has emerged over the centuries. (I say *he* advisedly, because it is a quite inescapable fact that before this century there were no truly front-rank women scientists. However the presence of top-notch woman scientists this century, such as

Marie Curie and Rosalind Franklin, shows this is a matter of opportunity not native ability).

Western civilisation is an individualistic civilisation. We prize our rights and liberties, we champion freedom of thought and toleration. Doing our own thing, getting to the top, fulfilling or realising yourself, becoming famous – all these ideals we take for granted in our culture. But they are not universal to mankind, and many societies have criticised (and still do criticise) such values as disruptive, selfish, or simply unhealthy. Individualism and science, however, have gone hand in hand. For Western science, particularly as it has developed over the last 500 years, has been a corpus of knowledge and ideas in constant change – even, to use Trotsky's term, in a state of 'permanent revolution'. And at the very centre of the march of science has been the individual scientist, dedicated, for whatever reasons or motives, to uprooting the authorised version of truth in his field, and substituting his new vision of things.

Such men have often been hyper-aware of their own powers, their daring, their radicalism. Kepler called his great work *Astronomia Nova*, the 'New Astronomy'; Galileo entitled his masterpiece *Discourses on Two New Sciences*; Newton looked back in later life – as a letter in Rupert Hall's chapter shows – on how he had revolutionised the direction of scientific inquiry during a year spent at his Lincolnshire home when his youthful studies in Cambridge were interrupted by the Plague. In China, during what we call the Middle Ages, Mandarin science built up a formidable array of learning (in many respects, well in advance of the West). But Chinese science grew stagnant, more concerned with the encyclopaedic goal of assembling and ordering what was known, than with adventuring into the unknown.

Chinese science, like Chinese culture, remained unindividualist: no Galileos or Darwins there. In the West, by contrast, the scientist emerged as the archetypal innovator, the man who was dedicated (as Francis Bacon phrased it) to the 'effecting of all things possible' within the 'advancement of learning'. Not all great scientists, however, were rampant iconoclasts like Galileo, enjoying a scrap with the authorities; and many were quite authoritarian themselves, as Newton became in his later years. But within Western science, the mark of greatness came to be the capacity to rethink old theories, to stand settled truths on their head, to see Nature through new spectacles. The discovery of the 'New World' of America was followed by a myriad of 'new worlds' in science – many revealed through new instruments, such as the microscope and the telescope, and some conceptualised by new ways of thinking.

The 'great man' theory of history is currently out of fashion – and with good reason. Francis Crick and James Watson discovered the double helical structure of DNA. But other groups in the scientific community were probably only weeks behind. It would also be shallow to think of the progress of science simply as the triumph of heroic minds over lesser ones. We have no real evidence that William Harvey was 'cleverer' than those who denied the circulation of the blood. Indeed, though Lavoisier got the chemistry of gases 'right', and Priestley got it 'wrong', many would argue that Priestley had by far the livelier mind. The scientists who ended up on the losing side weren't all dull stick-in-the-muds, or mere closed minds. Discarded scientific theories have always had plenty of facts and reasons to back them up, and often they seem far more reasonable than the theories which displace them. Hence there are great dangers in viewing history by hindsight, and assuming the foregone superiority of those theories – Galileo's, Newton's, Lavoisier's, Bohr's – which we endorse now. (After all, we can now see that Galileo, Lavoisier and so forth got so much wrong!).

And it would be equally silly to assume that great scientific theories popped out of the brains of scientific geniuses, rather like Pallas Athene arising out of the head of Zeus. Scientists are children of their times, and new science emerges – as did Napoleon – only when the time is ripe. Orthodox theories become hopelessly overloaded or over-sophisticated; new facts are discovered which generate anomalies that can't be absorbed in the existing conceptual frameworks. And not least, currents of thought outside science itself impress and impinge themselves upon the scientific mind. New outlooks in theology, philosophy, or in aesthetics, new perceptions of man and his mission, have always created fresh problems and opened up new perspectives for scientists. Thus one reason why Kepler could commit the heresy of denying the circular orbits of the planets was because the theological and quasi-mystical circles within which he moved, encouraged him to try to identify God's spiritual presence in the world. The Sun was surely emblematic of God. This helped Kepler to conceive of the sun's power (the *anima motrix*) propelling the planets in their orbits, *elliptical orbits*. Thereby a new scientific model – the ellipse, not the circle – could be embraced, thanks to what we might call non-scientific outlooks. The great scientist is not the man isolated from his age, but one so enmeshed with it that he changes it.

That is why it is important to focus upon the transformations made by individuals. For the work of individual scientists is more than the sum of the spirit of the age. No committee of the early Royal Society – a unique collection of talent – could have got round a table and written Newton's *Mathematical Principles of Natural Philosophy*. Doubtless, evolutionism was in the air in the nineteenth century, and many types of evolutionism were proposed, from Lamarck's 'inheritance of acquired characteristics' through to Herbert Spencer's comprehensive philosophy of the emergence of organic heterogeneity from homogeneity. But Charles Darwin's evolutionary theory (the survival of favoured variations by means of natural selection: the 'survival of the fittest') was distinctive. And it was distinctive because Darwin was a distinctive man. Versed in geology and palaeontology, steeped in regular natural science and natural theology, he had also spent five years on the voyage of the *Beagle* carrying out intensive natural history investigations in a multitude of exotic environments. (Not least, he had an evolutionist – Erasmus Darwin – for his grandfather!) Small wonder Charles Darwin came up with a distinct form of evolutionism – one most scientists now accept as approximately right.

Distinctive, but not unique. For at the very time that Darwin was perfecting his theory, the young Alfred Russel Wallace was formulating almost exactly the same idea, while suffering from a bout of malaria in Malaya. Surely, one might say, this fact – 'simultaneous discovery' – is a hammer blow against the special role of the individual: Darwin's very thoughts were being thought – almost telepathically – by another scientist. The answer is yes and no. True, Darwin wasn't the only person to come up with evolution by natural selection. But he was the only person with the right standing and right contacts ('pull') to get most of the scientific community and much of the general public to accept the theory. If only the unknown and obscure Wallace alone had come up with the theory, it would surely have remained as unknown as the genetic laws of the Czech monk, Gregor Mendel, remained for nearly half a century.

Science is thus individual talent. But it is also social relations. No more than anyone else can scientists breathe in a vacuum. Science operates – ever more so – in public and social dimensions; it involves vast financial outlays, apparatus, institutions, education, communication. Not least it involves politics. Scientists like to promote the view that science is outside or above politics (and thus that the 'abuse' of science is solely the fault of the politicians). This has never been true: for instance, the controversy between Priestley and Lavoisier was seen as a battle between English and French chemistry. And nowadays especially, the separation of science and politics is pure myth; after all, most fundamental science is directly funded by the state. Science aspires to be international, but cannot divest itself of national political pressures, especially in such sensitive fields as defence. Great scientists have recognised their responsibilities. Men of Kepler's century saw themselves duty-bound to God. Einstein, troubled over the Bomb, thought science responsible to mankind.

This book explores the unbounded capacity of scientific thinkers to extend our intellectual and technical horizons. Science's capacity to change mankind's moral and social perceptions is no less. So the great questions of science's future – how can it best be advanced? Can science be directed or must it be free to follow its nose? If we leave science free, how will it be accountable? – are questions for us all. Understanding science's past, will help us create its best future.

Roy Porter
Wellcome Institute for the History of Medicine

Aristotle (384–322 BC)

Ptolemy (C 100–170 AD)

Galileo Galilei (1564–1642)

Johannes Kepler (1571–1630)

ARISTOTLE:
The Theory and Practice of Science
Geoffrey Lloyd

The fertile crescent of the Near and Middle East was the cradle of civilisation. It was also the cradle of science. From Mesopotamia came the fundamentals of astronomy, theories of Earth and Heavens, and advances in technology. The Egyptians contributed much to measurement, surveying and to medicine. The Greeks assimilated these earlier traditions of knowledge and incorporated them into coherent theories of Nature, viewed as orderly, and open to human reason.

Aristotle was born in 384 BC, son of the court physician to Amyntas, King of Macedon, and, as a young man, was sent to Athens to study at the Academy under its founder, Plato. Some two decades later, after Plato's death, he left Athens and became tutor to Alexander, heir to the throne of Macedon and the man destined to conquer much of the known world. Upon Alexander's accession, Aristotle returned to Athens to found his own school, the Lyceum. Most of the known works of Aristotle are probably notes of lecture courses that he gave at the Lyceum. In 323 BC, after Alexander's death and in the face of anti-Macedonian feeling, Aristotle withdrew to Chalcis in Euboea where he died a year later.

For the originality and importance of his contributions to an amazing range of subjects Aristotle ranks as the most outstanding of all ancient Greek philosophers, perhaps of any philosopher of any time. In logic, physics and chemistry, zoology, psychology, ethics, sociology and even literary criticism, he either founded the subject as we know it or made a fundamental contribution to it. We shall be concentrating here on his work in natural science, but we should first mention briefly some of his other contributions since there is an important sense in which his work forms a unified whole, linked by a common methodology and by certain leading ideas.

In logic he may be said to have invented the study we call formal logic. He was the first to attempt systematically to analyse and classify valid and invalid arguments – the validity or invalidity of an argument being recognised to be independent of the truth or falsity of the premisses. In psychology or the philosophy of mind, where Plato had argued for a dualist view according to which the soul and the body are separate entities of totally different kinds, Aristotle maintained a monistic position closer to modern behaviourism. Just as seeing is the activity of the eyes, so in general the soul is the activity of the living organism, *its* vital capacities, not a separate entity inhabiting it like some 'ghost in the machine' as the modern philosopher, Ryle was to put it. In ethics Aristotle again reacted against Plato. He agreed with Plato that moral values are not just relative. According to the relativist there are no objective criteria for determining what is good or bad, right or wrong: if two people disagree there is no way of settling their disagreement, and both must be said to be right, each from his or her own point of view. Aristotle maintained, with Plato, that there are such objective criteria, but he rejected Plato's more extreme view that moral values are absolutes. Ethics, Aristotle insisted, is no exact science and questions to do with the good and the beautiful admit of some variation in the answers. In sociology Aristotle was the first to attempt the systematic, comparative, empirical analysis of societies, in his studies of the histories, especially the constitutional histories, of Greek states. Finally, in his discussion of literature and art, in the *Poetics*, he again argued against Plato. Where Plato had banished the poets as potentially subversive from his ideal state, Aristotle reinstated creative art, arguing strongly not just for the pleasure it gives but also for its educational value.

As that brief survey already illustrates, the starting-point of Aristotle's own work is often criticism of his predecessors, especially of Plato (under whom Aristotle studied for twenty years and with whom he nevertheless agreed, as we shall see, on certain fundamental issues). To review earlier work on a subject is, indeed, one of Aristotle's cardinal methodological principles. He insists that it is essential, before beginning an inquiry, to be clear about what the problem is, and one of the best ways of doing this is to analyse what earlier theorists had said – to discover where they had run into difficulties. Time and again in different branches of his work Aristotle first undertakes such a review of the most important ideas his predecessors had proposed. But he does this not, or not just, out of historical interest, but to examine their contributions to the substantive issues in each inquiry.

He often proceeds in this way in his natural science. Perhaps the chief problem that had exercised early Greek physicists was that of the ultimate constituents of physical substances – that is, element theory. A fifth-century BC philosopher called Empedocles – who was an extraordinary combination of physicist, spiritual leader and outright showman – had already put forward the view that the ultimate constituents are four elements, or 'roots' as he called them, earth, water, air and fire, all other substances being interpreted as compounds of these. One or other version of some such theory proved attractive to several other fifth- and fourth-century theorists (including Plato) before Aristotle modified and adapted it, but already in the late fifth century it faced stiff competition, especially from the first attempt at an atomic theory, the work of the philosophers Leucippus and Democritus. Although Aristotle's own version of the four-element theory owes much to Empedocles, he introduced certain modifications and above all supported it with powerful arguments against its chief rival, atomism.

The chief modification he introduced relates to the question of the ability of the four elements to turn into one another. Empedocles himself ruled that out, but Aristotle argued that it can and does take place. Thus 'water' turns to 'air' when it is heated, and 'air' again to fire when it burns: it should be explained that 'water' and 'air' are not thought of as chemically pure substances, but cover a variety of what we should call liquids and gases. (Again 'earth' covers an even wider range of solids – and although this lack of definition is one of the weaknesses of the four-element theory, one of its attractions is that three of the four simple bodies correspond, roughly, to matter in its solid, liquid and gaseous states.) Each of the four simple bodies was itself analysed, in Aristotle's theory, in terms of a pair of fundamental qualities: each was either hot or cold and either wet or dry. A great variety of ordinary observable physical changes and interactions, including some we should consider chemical combinations such as the alloying of two metals, could be interpreted in gross qualitative terms either as the exchange of fundamental qualities or as the interactions of bodies possessing them. Distinguishing between the mere juxtaposition of parts (as in a heap of grains of wheat and barley) and a combination in which a qualitative change takes place, Aristotle developed a theory that foreshadows in some features the modern distinction between a merely mechanical mixture and a chemical combination.

The early atomists had suggested a quite different style of physical theory which Aristotle discusses and sharply criticises.

They held that what existed fundamentally were atoms and the void alone. The atoms differ not in substance but merely in shape, position and arrangement, and they are assumed to be in constant motion in the void (one of Aristotle's complaints was that the atomists gave no account of motion, but simply assumed it – but Aristotle's own explanation of how the simple bodies move was in terms of their 'natural' tendencies). In their constant movement atoms collide with one another and some form compounds which correspond to the various kinds of substance we experience. Against this Aristotle argues that it is quite mistaken to attempt to reduce qualitative differentiae to quantitative ones. The problem, he insists, is to explain the tangible qualities of things: but hot and cold, wet and dry and so on cannot be seen as due merely to the essentially quantitative, mathematical differentiae of the variations in the shape, position and arrangement of the atoms. Although versions of atomism continued to be put forward after Aristotle, notably by the Epicureans, the dominant physical theory throughout antiquity – and indeed until the rise of modern chemistry – was Aristotle's.

The four Aristotelian simple bodies each have a natural movement, earth and water downwards, air and fire up, but 'downwards' is defined as towards, 'up' as away from, the centre of the universe deemed to coincide with the centre of the earth. Aristotle insists that bits of earth must always behave in the same way: imagining in a bold thought experiment what would happen if the earth itself was shifted from the centre, he argued that it would fall back and come to rest there. He believed that heavy objects fall downwards not in parallel straight lines, but in lines that converge at the centre of the earth, and he invokes this belief in one of the proofs he gives that the earth itself is spherical. But of course that belief about falling bodies could not, in his day, be confirmed empirically. However he also adduces, in support of his conclusion about the earth's shape, the convincing evidence of the changes in the positions of the stars observed at different latitudes and especially of those observed never to set, and further the shape of the shadow cast by the earth in eclipses of the moon. Given that lunar eclipses can occur at any point on the ecliptic, it follows that the earth itself is spherical. What is important about Aristotle's contribution here is that he provides the first extant *demonstration* of the sphericity of the earth. Others before him had maintained that theory, and others too had known that it is the earth that causes eclipses of the moon. But Aristotle now demonstrated a conclusion that directly conflicted with what 'common sense' assumed – that the earth was flat. Of course even after Aristotle some continued in their belief that the earth is flat. Yet his proof of its sphericity was a striking example of the power of science to demonstrate conclusions that completely overturned 'common sense' assumptions.

Aristotle himself was not a practising mathematician, but he was far from being as hostile to mathematics as he was later represented by, for example, Galileo. True, Aristotle favoured, as we have seen, a qualitative not a quantitative element theory. But first he made an important contribution to the philosophy of mathematics. In the philosophy of mathematics he argued (once again) for an alternative to Plato's view. Plato had held that what the mathematician studies are separate, transcendent, intelligible entities. Aristotle maintained, on the contrary, that there is no need to postulate a separate class of entities as the objects the mathematician studies. Mathematics is concerned, rather, with the mathematical properties of *physical* objects, though the mathematician studies these in abstraction from the other qualities that make them the physical objects they are.

In practice he appeals to mathematics or to the work of the mathematicians quite frequently in specific contexts in his natural science, for instance in astronomy. One notable occasion when he does so comes at the end of his discussion of the shape of the earth, where he gives the first recorded estimate of its size. He does not tell us the method used but we may presume that, like some later calculations, it was based on the difference in the altitude of stars observed at different latitudes. If the distance between two points on the same meridian is known together with the observed difference in the altitude of the same star, the circumference of the earth can be got by simple geometry. The result Aristotle records is 400 000 stades: we do not know which of several possible stades Aristotle used (they vary from about 157.5 metres to 148.8 metres), but in any case it is clear that it was an overestimate by nearly a factor of two. However the importance of the text, once again, lies in its showing *what was within the reach* of mathematical science.

Aristotle does not attempt to determine the sizes and distances of the main heavenly bodies (as some later Greek astronomers were to try to do), but his estimate of the size of the earth has the general consequence that, as he puts it, the earth is of no great size in relation to the sphere of the heavens. This did not shake Aristotle's confidence in its importance, for he still held that the earth occupied the central position in the universe. Yet he was faced with the difficulty of the evident disproportion between, on the one hand, the earth and the other elements with which we are familiar, and, on the other, the immensity of the heavens. If the heavens and the heavenly bodies were made, as some had believed, of fire, that would have consumed the other elements and the earth itself long ago. Nor again for similar reasons could they be made of air. Aristotle also knew very well that the observed motions of the stars, planets, sun and moon were, broadly, circular. All make a circuit of the heavens daily (it was not until after Aristotle that the possibility of explaining the same phenomena by ascribing an axial rotation to the earth was canvassed: for Aristotle himself the earth, as already noted, is at rest). Moreover, following the theory first developed by Plato's younger contemporary, the mathematician and astronomer Eudoxus, Aristotle used combinations of circular, concentric motions to explain the special movements of the planets, moon and sun along the ecliptic (the ecliptic being the path the sun itself takes in its yearly movement through the constellations of the zodiac). These extraordinarily swift and regular movements could not be unnatural: yet the natural motions of earth, water, air and fire are not circular, but in a straight line, 'downwards' or 'upwards', that is to, or from, the centre of the earth.

The size of the heavens, and the circular movements of the heavenly bodies, both pointed to the conclusion, Aristotle believed, that the heavens themselves must consist of a different element, aether, unlike the four simple bodies here on earth in being neither hot nor cold, neither wet nor dry (and so no threat to the balance of those opposites here on earth) and in having the unique property of moving naturally in a circle. This doctrine had momentous consequences – for which Aristotle has often been castigated – for it separated the heavens from the 'sublunary' region, the region below the moon, and seemed thereby to rule out the possibility of a unified natural science applicable to both heaven and earth. Yet it was in part Aristotle's correct perception of the tininess of the earth that led to this view. Moreover, in the heavens the movements of the planets, sun and moon were unimpeded and exact. It is often said that with the discovery of gravity Newton reunited the heavens with the earth. But Newtonian conditions of frictioness motion existed in the heavens for Aristotle, and so it might be truer to say that what Newton did was to allow the earth and terrestrial dynamics to be united with the heavens.

Aristotle had a theory of four causes that is best understood as a theory of the kinds of explanation of facts or events that can be given. The material cause concerns the question of what things are made of (one particular table may be made of wood, another of metal and glass). The formal cause picks out the essential as opposed to merely accidental properties of things (it is the shape and function that make a table what it is as opposed to something else, a bench or a bookcase for instance). The efficient cause relates to what brings change or movement about (the efficient cause of the table is the carpenter who made it). But in addition to these three, there is a fourth type of cause, or more correctly a fourth kind of explanation, to which Aristotle attaches the greatest importance throughout his natural science, and especially in his zoology. This is the so-called final cause, the good that an object, process or event serves. Take the lungs or the liver. It is essential for the biologist to understand not just what these organs are made of, but also their function and how they promote the well-being or the survival of the animals that possess them. Aristotle holds, for example, that the lungs serve to cool the region round the heart, and that some animals without lungs achieve a similar end – balancing body temperature – either by having a large body surface exposed to the air (as with many insects) or by the use of gills (which, Aristotle believes, take in water to effect a similar cooling). Many of the specific suggestions Aristotle makes about the functions of different parts of the body have been thought ridiculous. He was mocked, for instance, for saying that the eyebrows serve to keep the rain out of the eyes (though that is far from being an obviously foolish suggestion). But his search for final causes has been open to misunderstanding on several scores. First, although he often compares the activities of nature with those of human craftsmen – the carpenter, say, who makes a table or a chair to serve a particular function – he does not believe, in fact he denies, that nature consciously deliberates. Rather, there no intentionality is involved. Nevertheless his fundamental point is that when an organ, for instance, serves some good – for example in helping the animal to survive – then reference to that fact is part of the account to be given of the organ in question.

Secondly, he is well aware that other theorists had done without, or had denied, final causes. Empedocles and the atomists, especially, had often argued that things are as they are merely as the result of chance, the random interplay of what we might call mechanical causes, and they further held that it was animals that were well fitted to survive that did so (although they had no conception of the on-going evolution of natural species). Aristotle agrees that animals are in general well-adapted to survive but thinks of species as fixed. Moreover, it cannot be a matter of mere chance that species reproduce according to kind, for instance, and – as already noted – when something happens regularly and secures some good, then that good must be included in the explanation. The incisor teeth and the molars, to take another of his examples, did not just grow the way they did (neither when some animal first developed them, nor in animals that have them now): they have the shape they have in order more efficiently to cut, and to grind, the food. Final causes are in addition to, not instead of, other kinds of explanation, but they cannot be dispensed with. Thus it is true, but insufficient, to say that horns come from the surplus earthy material that horned animals consume: for they may provide a means of defence and so contribute to their survival.

Finally, it is not the case that Aristotle demands final causes for *everything* in nature. In some instances he himself allows that things are as they are, not to serve some function, but merely because they happen to be so: they *are* just the result of mechanical or material processes. One of Aristotle's examples is the colour of the eyes. That a particular individual has blue, or hazel, eyes serves no particular good. But contrast what you have to say about having eyes at all.

Despite the waywardness of some of Aristotle's specific suggestions, teleology – the search for final causes – served as an important heuristic principle, that is to say it directed the biologist to ask the question of the possible function that a part, an organ or a process fulfils. Moreover, it provided Aristotle with his chief motivation for a detailed empirical programme of research in zoology. Plato too had believed that the cosmos manifests the workings of design. But Aristotle investigated this systematically in various areas of his natural science but especially in his study of animals, the first comprehensive comparative zoology. Moreover, in this field he brings to bear a powerful new technique of research, the method of dissection, which although it had been used occasionally, had never before been employed so extensively on animals (it was left to the later biologists Herophilus and Erasistratus to extend the method also to humans). Against the Platonists, who had little time for detailed empirical research, Aristotle feels he has to justify his procedures and this he does by insisting that he is investigating not just the material causes but also, and more especially, form and finality – for the good and the beautiful are to be found in even the humblest of living creatures. Aristotle certainly emphasises the importance of collecting the data before proceeding to their explanation: but it must be stressed that nowhere is data-collection an end in itself, it always serves as a preliminary to an account of the causes at work.

One of the principal reasons for the extraordinary ascendancy Aristotle had on subsequent Greek thought lies in the completeness of his programme. His theory of causes specifies the questions to be investigated, and in one field of natural science after another – as in many other branches of inquiry – he provided clear analyses of the difficulties and often coherent and plausible resolutions of them. But although there is a sense, I said, in which his work forms a unified whole, linked by a common methodology and a preoccupation with forms, or essences, and final causes, his thought was not the closed and dogmatic system that some later took it to be. In the Middle Ages he was the 'master of those who know' and the reinterpretation of his ideas by Christian theologians, such as Aquinas, led to the identification of Aristotelianism as *the* dogmatic school philosophy.

But there was a second, more sinister, reason for the popularity of Aristotle, at least in certain quarters. His system proved difficult to undermine, in part, because it presented an overall integrated view of the cosmos, but that picture was essentially a hierarchical one in which every kind of natural species had its proper place. The heavenly bodies – divine manifestations of supreme regularity – man, the other animals, plants, down to the lifeless substances and the simple bodies themselves were all conceived as ordered in a hierarchical whole, ranging from the most, to the least, perfect. This was a system that reflected, and in turn was used to justify, the authoritarian traits of ancient Greek society: Aristotle personally held that slavery was a natural institution (although some are slaves only by accident) and that the female sex is essentially inferior to the male. But in later ages too it was invoked to give some kind of cosmological warrant to the perpetuation of hierarchical political arrangements, the right of some men to rule, while others were there to be ruled.

While the authoritarian tendencies in Aristotle's thought are undeniable, to represent him as a dogmatist is a drastically one-sided view. Although he naturally seeks answers when answers can be given, it is often in the formulation of the questions that he is at his most acute. Moreover, he repeatedly stresses, that certain questions remain open, that he is not yet in a position to resolve the difficulties, that more research on particular points has to be undertaken. His ideas on many topics changed and developed during the course of his life and a certain tentativeness characterises his approach to many important issues.

These aspects of his philosophical method would nowadays be considered among his most durable legacies. While most of his specific scientific theories have, naturally enough, long ago been superseded, the concept of science itself that he put forward has more than just a historical importance. I have already remarked on his recognition of the importance of empirical research, which he did not just advocate but also practised. Data-collection – and the review of earlier theories to identify where the difficulties lie – are necessary preliminaries in any inquiry, but as I said, they are *just* preliminaries, for the goal is always *understanding*. For that, what is needed is not just valid inferences from true premises: the premises must contain the *explanations* of the conclusions. Ideally a mature science should be able to *demonstrate* the truths it reveals, and even if not many sciences were in a position to do so in Aristotle's day, he identified the task.

In the early stages of Greek science many attempts had been made to answer far-reaching questions to do with matter, with the nature of change, whether the cosmos is eternal or created, with the nature of the soul or mind, and so on. In some branches of inquiry evident successes had been scored – as when early Pythagorean investigators discovered the numerical ratios (1:2, 2:3, 3:4) in which the simple concords of octave, fifth or fourth can be expressed, or when the Hippocratic doctors insisted that all diseases have a natural cause. Yet so far as we can judge from our admittedly often fragmentary sources, such successes were not the outcome of a clearly defined programme or methodology. What Aristotle – building here on Plato's work especially – was the first to do was to provide a clear analysis of the notions of understanding, explanation and demonstration: he acknowledged the greater, or less, exactness of inquiries depending on their degree of abstraction and he insisted throughout on a certain self-consciousness in method. Even those who disagreed with his view-point could not fail to be impressed by his extraordinary confidence in the scientific enterprise. While philosophy of science, like science itself, has of course moved on from where Aristotle left it, it is in this area especially that his ideas remain influential today. When modern philosophers of science debate teleology or essentialism or realism, often in admittedly technical and sophisticated language, the positions and arguments that Aristotle developed when he discussed final causes, forms and essences, and the truths that science reveals, are of more than merely historical interest and continue to form the basis of new suggestions for the clarification of far-reaching problems.

PTOLEMY:
The Synthesis of Ancient Astronomy
John North

The great age of Greek civilisation – of Athenian democracy – is generally taken to be the fifth and fourth centuries BC. But the tradition of Greek science continued to flourish for long after, and especially outside Greece itself. Euclid, the great geometrician, lived in Alexandria around 300 BC and Archimedes, the founder of statics and hydrostatics, lived in Sicily in the third century BC.

Ptolemy was the greatest astronomer within the Greek tradition. He lived in Alexandria in the second century AD and has been immortalised as the author of one work which synthesised all the Ancients knew of practical, computational astronomy, and another which did much the same for geography. Thanks to his writings, the Medievals were remarkably good at astronomical prediction. Ptolemaic astronomy contained, however, two great flaws. First, Ptolemy put the Earth at the centre of the planetary system. Second, he divorced the calculational side of astronomy from the physical. These deficiencies were to survive well over a thousand years until the insights of Copernicus and Kepler.

Ptolemy – Claudius Ptolemaeus – was the most influential astronomer of antiquity, and in many ways the greatest. Well into the seventeenth century his name was to astronomy what Euclid's was to geometry and Galen's to medicine. Whilst his achievements were forgotten for a time in western Europe, his fame and his methods alike had been preserved in the East, in particular in Islam. Ptolemy's finest astronomical work, the collection of thirteen books known as the *Almagest*, is scientific writing at its best. Copernicus took it as a model, even though he was undermining its assumption that the Earth is central to the universe – a premiss of almost all ancient astronomy. Ptolemy's reputation outside astronomy proper lasted even longer, for he wrote what became the fundamental reference work of astrology – the *Tetrabiblos*. He is remembered too for his compendious *Geography*, which records much invaluable information about the ancient world that is to be

found nowhere else. In it he established mathematical foundations for scientific map-making. No less important are his writings
on optics. 'Ptolemy the divine', he was called by Hephaiston of
Thebes, not entirely without reason.

It is hard, however to make an idol out of a man about whose
personal life we know so little. We do not know exactly where or
when he was born or died. He lived and worked in or near
Alexandria. It has been said that he was born in Ptolemais Her-
mion, and this must have been close to AD 100. His name 'Claudius' tells us that he held Roman citizenship, and 'Ptolemaeus'
that he was either of Greek extraction or a Hellenised Egyptian –
probably the former judging by his condescending way of
referring to the Egyptians. According to the Arab astronomer
Abû'l Wafâ' – who was writing eight centuries later, however – he
lived to the age of seventy-eight, and another writer tells us that he
lived into the reign of Marcus Aurelius. His astronomical observations date from the period AD 127 to 151. In short, Ptolemy lived
from about AD 100 to about 175 in that part of a province of Egypt
which, whilst it was Roman by conquest was Greek in its thought
and its traditions. Greek Alexandria, in Ptolemy's time, was four
centuries old. It had been a Roman province only since 30 BC, and
the *lingua franca* was still Greek. Power in Alexandria was in the
hands of Macedonians and Greeks, but races and cultures were
there mingled to an unusual degree. Much eastern learning was
brought to the Mediterranean from beyond the limits of the
Egyptian empire, and the part of this that most concerns us is an
extraordinarily rich astronomical tradition deriving from the
Babylonians and the Seleucid Greeks of the Babylonian region.

The astronomy of the Babylonians was essentially *arithmetical*
in character, based on a powerful sexagesimal arithmetic – a
system based on the number sixty, traces of which we still preserve
in our minutes and seconds of an hour. They made little or no use
of geometrical models of the heavens, which were being brilliantly
exploited by the Greeks of a generation before Alexander, men
such as Eudoxus, Callippus, and Aristotle – Alexander's own
tutor. The most important Alexandrian institutions from an
astronomer's point of view were two: the Museum and the
Library. The Museum – as its name should suggest – was a temple
of the nine muses. It had a staff of learned men, who held property
in common, presided over by a priest. It seems to have had space
set aside for astronomical observation – astronomy, after all, had
Urania as its muse. Even though the Museum had lost its earlier
importance as a centre of learning by Ptolemy's time, the writings

of its members were preserved in that other great Alexandrian institution, the Library. The works were of course not codices, books as we know them, but rolls of papyrus, difficult to catalogue and easy to damage. In his *Tetrabiblos*, Ptolemy complains about an ancient example he has been reading with great difficulty, and tells us how much better preserved it was on the inside of the roll. An astronomer in those days had to be a historian, a palaeographer, a textual critic, and philologist. He was truly fortunate if in addition he had access to the spiritual and material resources of Alexandria. Of course he had to be a mathematician and scientist too. Ptolemy was evidently all of these things.

When we speak of 'Greek mathematics' we refer not to a tightly knit intellectual movement but to the fragmentary writings of a relatively small number of geographically scattered mathematicians from a wide spread of centuries. No doubt to the front of our minds are just three or four men – say Euclid, Archimedes, Apollonius, and Diophantus. Likewise with 'Greek astronomy': it is in order here to mention briefly again just four men, all of them of great historical importance. (A fifth would have to be Aristotle, for his having bonded together physics and the astronomy of Eudoxus.)

Eudoxus, who lived five centuries before Ptolemy, offered a brilliant geometrical explanation of how it comes about that the planets behave in the odd ways they do. They vary their speeds as they move against the background of 'fixed' stars, and even move in a retrograde direction from time to time. The classic Greek problem had been to explain all this in terms of motions that were both *uniform* and *circular*. Eudoxus was from Cnidus, and spent much time in Athens, where he was acquainted with Plato. The planetary theory of Eudoxus was based on a complex system of concentric spheres – rather like the shells of an onion, but pivoted around a variety of poles.

A second astronomer worthy of mention here was Aristarchus, celebrated as the first with a Sun-centred theory. Although it was only descriptive in character, how different history might have been had the idea caught on! He was of Samos, near Miletus, and probably studied in Alexandria, but he did so four centuries before Ptolemy, more than the span of time that separates us from Galileo.

Two other 'Greek' astronomers must be added to the list. Apollonius, more eminent as a mathematician, was of Perga, a small Greek city in Asia Minor. He lived more than three centuries earlier than Ptolemy, and yet only through Ptolemy's

Almagest do we know of his achievements in astronomy. Apollonius made use of an idea less difficult but far more fruitful than the 'homocentric' hypothesis of Eudoxus: this was the *epicyclic* model, which we shall find Ptolemy adopting and improving. The planet is supposed to move in a circular orbit, but the centre of that circle in turn moves round the Earth. By combining circular motions in this way, it is possible to explain very simply the retrograde motions of the planets. Apollonius proved the geometrical equivalence of a theory with a simple epicycle and a theory with a simple eccentric – of which more shortly – and gave an admirable analysis of the problem of determining the stationary points on a planet's path.

The last important Greek precursor of Ptolemy was Hipparchus, born in Nicaea in north-west Asia Minor. He made his observations somewhat less than a century and a half before the Christian era, in Rhodes. Only one relatively unimportant work of his survives, and yet from the respect in which Ptolemy held him, and from the details of his work that the *Almagest* transmits, we know that he was of the stature of Ptolemy himself. Once more, however, the time interval must not be overlooked: it as as though all that posterity knew of the writings of Newton was what could be learned from an ordinary mathematical text-book of the present day. (Oddly enough, just as Newton had his portrait on a recent pound note, so Hipparchus had his on Nicaean coins, three or four centuries after his death.)

Hipparchus wrote on mathematical methods for astronomical use. He drew up tables of chords (related to our tables of sines) and was the first to put trigonometry on a general footing. Indeed, without stretching unduly the meanings of words, we may say that he invented trigonometry – and not the simple sort we learn in our high schools, but trigonometry as it applies to triangles composed of arcs of great circles on a sphere. Ptolemy made great use of his techniques, and elaborated upon them to great effect.

Hipparchus devised geometrical models to account better than ever before for the motions of the Sun and Moon; and Ptolemy improved on them. Ptolemy drew heavily on Hipparchus' star catalogue, which had contained perhaps 850 stars, although he made use of the catalogues of others, too, and Ptolemy's final list included 1022 stars, with their coordinates. His great forerunner had in fact, through careful observations on star positions, discovered one of the most subtle of celestial movements, that of the stars against the equinoxes and solstices. (The vernal equinox, where the Sun crosses the celestial equator at the onset of spring, is a reference point in the heavens which ancient astronomers regarded as absolute. Its function as a starting point for longitudes resembles that of the Greenwich meridian on Earth.) The extremely slow motion of the stars was put at one degree per century, which is a little on the low side, but the result was truly remarkable, and Ptolemy accepted it.

Hipparchus had made a mathematical analysis of the problem of the time it takes arcs of the ecliptic to rise. (The ecliptic is the path of the Sun against the background of stars. Problems of rising time have to do with the length of daylight and with time-keeping generally, not to mention mathematical astrology.) He had used Babylonian arithmetical methods, but evidently supplemented them with his own trigonometrical techniques. Before his time, Greek astronomy had been a science more or less geometrical in character. Hipparchus turned it into a science that was not only quantitative but extremely precise – by far the most precise empirical science before modern times. He inherited the precision of the Babylonians, and he came at a cross-roads in history. In Ptolemy, we find that the old Babylonian arithmetical methods were more or less dispensed with, and that trigonometrical methods prevailed.

Insofar as we can judge from the imperfect evidence, the story was the same throughout most of astronomy: Ptolemy took from Hipparchus, but always improved upon what he had taken. As a consequence of this intimate relationship between the work of two men, a strange historical tradition has grown up that would have us believe that everything in the *Almagest* not claimed as his own – and perhaps even things that he did claim – was taken by Ptolemy from Hipparchus. The matter is historically complicated, but my own view is that Ptolemy was rather more open than were most early scholars, in acknowledging debts, and that he is not the sinister plagiarist some would make him out to be. One of the strongest arguments used by the sceptics is that Ptolemy took from Hipparchus a length of the year particularly poor by their standards (about six minutes too long) and yet presented observations that seemed to confirm it. The probable explanation is that he simply selected from his records those that would confirm a figure in which he had too much confidence. The phenomenon is not unknown elsewhere in science!

The Sun does not move round the sky uniformly in the course of the year, but its changes in speed are slight, and can be accounted for extremely well if we simply suppose that the Sun is indeed moving at constant speed round a circle, but that our Earth is displaced slightly from the centre of the circle. We are 'off-centre', and accordingly we say that the Sun moves on an 'eccentric'. In the case of the Moon and planets, an eccentric will not suffice – the motions are too complicated. Hipparchus used an epicyclic model for the Moon, and to determine the motions and geometrical proportions he made use of a long series of Babylonian eclipse records as well as observations of his own. In consequence, he was able to predict eclipses himself. It is a surprising fact that Ptolemy was the first for nearly three centuries to exploit Hipparchus' insight, applying it to the planets as well. He tells us that Hipparchus was content to show that existing theories of planetary motion did not conform with observation. The most he could find in Hipparchus was a body of data concerning planetary velocities – and we now know that they were of Babylonian origin.

When he came to examine Hipparchus' theory of the Moon closely, Ptolemy found that it worked well at opposition and conjunction (full and new moons; this was not surprising, since it had been based on eclipse observations), but not so well at first and last quarter, when the Moon was 90 degrees from the Sun, measuring along the ecliptic. His solution was a remarkable example of how to adapt the details of a theory to account for a set of accurate observations. He not only put the centre of the Moon's deferent circle away from the Earth (so that the Moon's was an 'eccentric deferent'), but proposed the hypothesis that this centre moves round *another* small circle. Associated with each circle there is a specific velocity, the parameters of the model (circle diameters, and velocities) being ultimately settled on an empirical basis. Ptolemy's methods for deriving the parameters were as important in their way as the model itself. When it was completed, he had created a theory of the lunar motions which was unsurpassed for fourteen centuries. It had one drawback, to be sure: it was designed only to predict the *positions* of the Moon, and not the proximity of the Moon to the Earth. This factor affects the apparent *size* of the Moon, of course. Ptolemy never mentioned the difficulty, but his theory has the consequence that the Moon should almost double in angular size every month.

This all touches upon an interesting question: would this palpably false prediction have been at all disturbing to Ptolemy? It has been said that he was interested only in positions, and that he was not in the slightest degree concerned with actual orbits in three dimensions. He wanted only, on this account, to 'save the appearances', the appearances being those from the Earth, and not from some position in outer space.

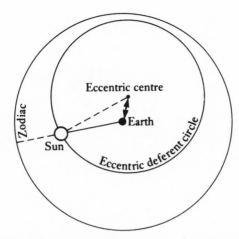

A simple eccentric model, particularly successful for the Sun. If the Sun moves at uniform speed round the eccentric circle, its velocity as seen from Earth will seem to vary according to its position.

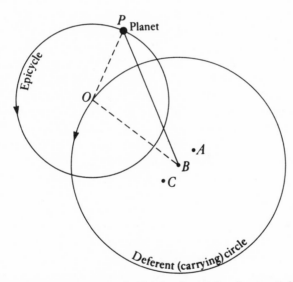

The simplest epicycle model takes the Earth to be at 'B' – the centre of the deferent (carrying) circle. The point where 'BP' meets the Zodiac (not shown) is the apparent position of the planet 'P', and when the planet is in the neighbourhood of Earth it will sometimes appear to move backwards along the Zodiac.

Many modifications can be made to this model. The Earth can, for example, be taken to be eccentric to the deferent circle and placed, say, at 'A'. Ptolemy used this device to great effect to account for planetary motions. In another refinement he made the angular motion of 'O', round the deferent, uniform around 'C', which he called the 'equant centre', instead of around the Earth at the true centre.

33

This is the traditional historical view of the matter, but there is good reason for thinking that Ptolemy did want to represent reality as closely as possible. If you read the *Almagest*, you will find that each planet is treated as a separate entity, with its eccentric deferent and its epicycle delineated, and the parameters of its model derived, but all more or less independently of the other planets. To a physicist, who expects the universe to comprise a unified system, with universal laws, this seems unsatisfactory, even though it may save the appearances. Aristotle, five centuries before, had left to posterity a model with concentric spheres, interrelated to the extent that each was supposed to act on the sphere next below it, so that the entire universe was activated from the First Mover, the Primum Mobile, at the outside. No matter that Aristotle did not account for the precise movements of the Sun, Moon, and planets; at least, there was a physical plausibility in his system that appealed to scholars for two millennia. What we now know, from a recently discovered portion of a work of Ptolemy (the *Planetary Hypotheses*) is that he tried hard to fit his eccentrics and epicycles into a model of the Aristotelian sort, so that he too could achieve physical plausibility for his system. The old view of Ptolemy is simply wrong.

There were other ways in which Ptolemy improved the positional predictive power of the astronomy he had inherited. He introduced a device known as the 'equant', for instance, to help in the fine-tuning of his theories. He accounted rather well for the latitudinal movements of the planets. (They do not stick closely to the ecliptic, but wander away to various angular distances from it.) The problems he faced were considerable, and we can now see why. The planetary system is Sun-centred, and the different planes of the planets' orbits pass through the Sun, not the Earth; but it is the Earth that is the centre of observation of the planetary latitudes. Copernican insights were needed before the problem of latitude could be made more tractable, and the marvel is that Ptolemy made such progress as he did.

The *Almagest* is as complete a manual of astronomy as could have been expected in the ancient world, and of course it contained descriptions of some of Ptolemy's instruments. There is a description of a celestial globe, for instance, depicting the stars in their constellations. There is also a description of what is called an astrolabe, not the instrument comprising a set of flat metal discs (like the planispheres one can still buy) that later usually went by the name 'astrolabe', but a set of pivoted rings with movements in three dimensions. His astrolabe had a didactic purpose – it lent itself to easy visualisation of the principal circles of the celestial sphere – but one of its rings was fitted with sighting holes, so that it could be used for observation.

The other sort of astrolabe (the planispheric sort) is an object of an entirely different order of mathematical sophistication. It is used for sighting the angular altitudes of stars, Sun, Moon, or planets, and for subsequently deducing such things as times, or geographical directions, or even for casting horoscopes. It required its maker to map the heavens on a flat surface, using what is known as stereographic projection – which is thought to have been another of Hipparchus' inventions. Ptolemy wrote a short but impressive work, *The Planisphere*, on the problem of representing circles on a sphere as circles on a plane. It is just possible that Ptolemy knew of an instrument resembling the medieval (plane) astrolabe, a disc-like object a few inches across, for there were elaborate water-clocks in his day, driving very large astrolabe-dials. Another work of Ptolemy's on a related subject is the *Analemma*, surviving only in a medieval Latin translation from the Greek. It is an account of the complex practical geometry needed for designing certain types of sundial.

Hipparchus had pointed out that before reasonable world-maps could be drawn, it would be necessary to discover by observation the latitudes and longitudes of all places to be represented. This was a task for the astronomer, but it was far from completed, even for the Mediterranean, in Ptolemy's day. Marinus of Tyre, who lived shortly before him, had tried to supplement what little was known with the approximate distances between towns that were deduced from journey times. In his *Geography*, Ptolemy took over all this material, and combined with it his own system of mathematical geography, which was all the richer for his experience with the general problem of plane projections of the sphere. He wished to show only the *known* world on his maps, which therefore required very special kinds of projection. He used two and both were extremely ingenious. The second is more or less peculiar to Ptolemy, while the first is virtually what we now know as a 'conical' projection. Ptolemy discussed the size of the Earth – preferring the estimate of the Greek philosopher Posidonius to the more accurate figure of Eratosthenes – and the size of the known world. He listed in tabular form about eight thousand places, mountains, rivers, seas, and so on. Not until the last century was it appreciated now uncannily accurate was Ptolemy on the source of the Nile. He believed that Europe and Asia together stretched half-way round the world – in fact they cover only about

130 degrees of longitude. His error, which was perfectly excusable, must have been something of an incentive to Columbus and the early circumnavigators.

It is often said that Ptolemy was a mathematical geographer without the slightest interest in human geography, but this is wrong, as can be seen from his astrological work, *Tetrabiblos*. There he draws a fundamental distinction between two sorts of prognostication by astronomical means, one universal, concerning races, countries, and cities, the other specific, and concerning individual human beings. It is almost always the second that comes to mind today, when anyone speaks of astrology. This more specific sort of astrology (genethlialogy) was given by the Greeks to Rome and thence to medieval and Renaissance Europe.

Astrology, however, had its roots in the first sort of practice, especially in the hands of the Babylonians of the Seleucid period; and when individuals were, at this early period, the subject of the art, it was usually as princely representatives of nations and peoples. The *Tetrabiblos* has much in the former category, something that we might perhaps more generously regard as 'astrological ethnography'.

Ptolemy begins at a very general level, with the straight-haired, tall, white-faced, well-nourished savages from northern parallels, for instance, and relates their conditions to their geographical situation. He is not content with generalities, though, and, in the end, few nations of the world escape his analysis – thus the British are described as a liberty-loving, industrious, and warlike people, with qualities of leadership, magnanimous, but without sexual passion. (Heaven only knows what books in the Library he had been reading.)

The *Tetrabiblos* is a mine of information on daily Alexandrian life, custom, and attitudes, but that is not, of course, why it was written. We see something of its author's racial prejudice, and we may discern a certain scepticism in references to 'uncultured, mad, easily frightened, superstitious frequenters of shrines and public confessors of ailments', and yet at the same time we have those long strings of association of planetary configurations with their human consequences that to modern eyes look more like the free association of ideas than genuine science. (This is most true of the doctrine he inherited from earlier writers.) There is, even so, a genuinely scientific motive at play in all this, a serious ambition to expound a systematic theory of human affairs – and one of meteorological phenomena too. It is clear that Ptolemy – unlike most other astrologers – tried to produce a coherent *physi-*

cal theory of astral influence. If he and others failed, this does not mean that the attempt was any the less valiant.

Ptolemy was known to posterity chiefly through books we have already mentioned – the *Almagest*, the *Geography*, and the *Tetrabiblos*. His chief work of physics, the *Optics*, is now preserved only in a Latin version of an Arabic translation. In it he covers what had become conventional doctrine: how objects appear in size and shape, how they are to be represented in painting – a problem related to that of map-projection – and the study of curved and plane mirrors. He contributed something new, however, in the form of physical and psychological explanations of vision and a theory of the refraction of light. Throughout the *Optics* there is ample evidence of his delight in experiment, whether in double vision, the merging of colours on a spinning disc, or the phenomenon of the Moon's appearing larger when near the horizon. (He related the last to the exertion used to elevate the direction of sight, and was foreshadowing a theory which has been partially accepted in modern times.) He found no law of refraction, if by that we mean an explicit mathematical relationship between the angles of the incident and the refracted ray, but he did present a table of angles in which a certain (mistaken) relationship was implicit. On the face of things, this table is a simple record of experimental measurements, and there is no reason to doubt that he did make measurements. The table is adjusted, however, to conform to a simple (quadratic) rule, and it is quite plain that the adjustment was second nature to a man trained in the tabulation of astronomical data. There is nothing shameful in making this sort of adjustment, which is exactly what we do when we formulate laws in mathematical form – although we should distinguish much more decisively between experimental findings and law. Alas, Ptolemy did not, and represented impossible readings (beyond the so-called critical angle). No matter: this table was to all intents and purposes the first genuinely empirical law of refraction. For small angles it was close to the sine law, the version we now accept, and which was found by Thomas Harriot in the seventeenth century – for all that, in Britain, we call it 'Snell's law'.

Ptolemy wrote a number of short tracts not discussed here, some of them known only by hearsay. He wrote on the balance, on music, on mechanics, and on the proof of the parallel postulate, for instance. The *Almagest* contains astronomical tables, but they are scattered throughout the book, and one of his tasks was to prepare a set of *Handy Tables* for the use of computers. As far as his astronomy was concerned, however, the *Almagest* was his supreme achievement, and it became a standard of reference from the very first. It was translated into Syriac, and (often) into Arabic, beginning in about AD 800. The astronomers of Islam did not accept the *Almagest* slavishly, but improved it in many small ways, and in one very great respect – in regard to the faulty solar theory. They occasionally tried to effect radical modifications to the basic models, even reverting to models resembling that of Eudoxus, but in the end, medieval astronomers as a whole returned to Ptolemy. The work entered western Europe in two versions, one translated from the Greek, the other from the Arabic, both in the twelfth century. The amorous Nicholas in Chaucer's *Miller's Tale* had a copy of the *Almagest* at the head of his bed. For roughly four centuries, western astronomers gave all their energies to mastering it, writing digests – which were read far more often than the full version – and finding new ways of manipulating it; but they were reluctant to modify more than the parameters it contained, and this they usually did with the help of Islamic writings. Even the work that finally led to the overthrow of the Ptolemaic system, Copernicus' *On the Revolutions*, could pay it no more flattering compliment than that of following exactly the same pattern of exposition, and incorporating very much of its substance.

The *Tetrabiblos* had a similar history, but the impressionistic and uncontrolled character of astrology meant that the book had many more rivals. The *Optics* had an influence that was profound but oblique, and that operated through three or four derivative writers – such as Ibn al-Haitham in the eleventh century (known in the West as Alhazen), and Witelo and Roger Bacon in the thirteenth century. The *Geography* was turned into Arabic in the ninth century, and was soon replaced by improved geographical works. It reached Europe late, and from the Greek; its popularity had much to do with the humanistic revival of Greek learning – here were Greek manuscripts all the more appreciated for not being overlaid with Arab learning – and may be gauged by the fact that there were no fewer than seven printed editions before 1500.

There was not a single edition of the *Almagest* in the same period: digests had captured the market. We know little of Ptolemy the man. His works are his monument; and for all that the sixteenth century thought otherwise, the *Almagest* is the finest monument of all.

GALILEO GALILEI:
'Modern' Science
Colin Ronan

Galileo Galilei (1564–1642) not only relished his battles with the Church, but also enjoyed quarrelling with the latter-day Aristotelians, heirs to the ancient science handed down through Islamic culture, to the Middle Ages. Galileo pictured himself as a heroic protagonist of the 'New Science' against the dogmatic absurdities of the 'Old Science'.

Galileo's Europe was extending its boundaries in many directions. The new world of America had been discovered. Knowledge, travel, trade and industry (so it was believed) would all advance together. And machines became a symbol of this alliance. Machines such as looms, clocks and guns improved in design; new scientific apparatus was invented (e.g. the air pump and telescope); and Nature itself was thought of as a wonderfully functioning machine, designed by God, the Great Mechanic. In Galileo, a practical interest in machines and the science of mechanics advanced together.

In 1564 at Pisa in the north-west of Italy, Galileo was born – a true son of the Renaissance then sweeping the Western world and one who was to leave an indelible mark on its development.

The Renaissance embraced the sciences as well as the arts, giving rise to a 'scientific revolution' which was to cast aside the previous customary dependence on ancient authority. On most matters the majority of western Europeans looked to the Christian faith for guidance, but science tradition required that they turn back to the teaching of ancient Greece. Here the primary authority was Aristotle, who had lived some nineteen centuries earlier. By and large, though, it was an Aristotle purged of his 'pagan' attitudes and interpreted according to Christian tradition. Indeed, this approach had become so entrenched in Italy, that to doubt Aristotle's science could bring one perilously close to a charge of questioning Christian authority. Galileo, then, was born into what was to prove an intellectual minefield.

First educated privately at home in Pisa, when later the family moved to Florence, Galileo was sent away to school at the Jesuit Monastery in Vallombrosa. Here, at the age of fourteen, he was entered as a novice of the Jesuit order, but his father strongly objected and immediately moved his son back to Florence. Yet it is clear that Galileo's three years at Vallombrosa had a deep effect on him; he remained a staunch Catholic for the rest of his days, and this in spite of his later disagreements with the Church authorities.

When he became seventeen, Galileo was sent to the University of Pisa. At his father's insistence, he was enrolled as a medical student though his interest in the subject was marginal; he much preferred the mathematical sciences. So, in spite of parental opposition, Galileo had private tuition in mathematics and made rapid progress, though in 1585 when he left the university, he did so without a degree – a not uncommon occurrence at that time.

While at Pisa, it seems that Galileo discovered the isochronism of the pendulum (the fact that it takes the same time to complete every swing, large or small). The story is that he found this out during a tedious sermon in Pisa Cathedral by timing the swings of a chandelier against his pulse. And it was while at the University that he also earned himself the nickname 'Il Attaccabrighe' ('The Wrangler'); this because he would argue with his lecturers, questioning their blind acceptance of the opinions of Aristotle and of the Greek physician Galen, whose teaching was still the foundation of anatomical study. Experimental evidence should be the touchstone of science, he claimed, not ancient authority, however eminent.

In asserting his independence of mind, the young Galileo was echoing his father's footsteps, for Vincenzo Galilei, a noted professional musician, had revolutionary ideas about music and was not backward in making them known.

On his return to Florence, Galileo took in private pupils and pursued his own mathematical studies, paying particular attention to the work of Archimedes, whose corkscrew-like device for raising water – the 'Archimedean screw' – is in use in some parts of the world even today. In 1586 he wrote a small tract *La bilancetta*, retelling how Archimedes discovered buoyancy when he was determining whether or not the new crown of his patron and friend, King Heiron, was made of pure gold. Galileo's text also described an improved design of hydrostatic balance for determining the density of liquids. As the years passed so the young Galileo became known and respected by Florentine intellectuals, and in 1589 was appointed professor of mathematics at the nearby University of Pisa.

Though friendly with the professor of medicine and with the philosopher Jacopo Mazzoni, Galileo did not get on well with other academic colleagues. This was due in part to his attacks on Aristotle's physics which was still accepted in university circles. Indeed, if the story is true that he did drop different weights of the same material from the famous Leaning Tower, his aim would certainly not have been to discover the fact that they hit the ground at the same time – that was already known – but to demonstrate that Aristotle was wrong to teach that the heavier would fall faster. Yet again he was keen to emphasize the folly of relying uncritically on authority, unsupported by experimental evidence.

But this was not all. He wrote a biting satire about the ordinance that academics must wear gowns at all times, an exposé of pompous regulations which allowed him to exercise his flair for demolishing his opponents by ridicule. And as if this were not enough, Galileo had the temerity to criticise severely a scheme for dredging nearby Leghorn harbour proposed by the rich and powerful Giovanni de' Medici. Neither was calculated to endear him to the authorities, and it is hardly surprising that when his three-year contract was up, the appointment was not renewed.

Yet if Galileo failed to get on with the authorities at Pisa, his time there was not wasted. He thought long and deeply about a new approach to physics, and wrote an untitled book known now as *De Motu* ('On Motion'). Besides developing a theory about falling bodies, Galileo made his first attack on Aristotle's ideas of motion, and began to formulate his own theory. This, as we shall see, was to be published in fuller form at the end of his life.

In 1591 Vincenzo Galilei died and the burden of providing for the family fell on Galileo's shoulders as eldest son. With the help of his friend, the Marquis Guidobaldo del Monte, who had been instrumental in getting him appointed to the chair of mathematics at Pisa, Galileo now obtained a similar post at the University of Padua. Here was an environment more to Galileo's taste. Situated in the province of Venice, noted for its freedom and independence of thought, he could now breathe the air of intellectual freedom. Gone were the shackles of narrow Aristotelianism which had bedevilled him at Pisa.

In Padua Galileo was made welcome by the Pinelli circle, given the run of Gianvincenzo Pinelli's library of some 80 000 volumes, and it was here that he met Paolo Sarpi, some twelve years his senior. A man of broad interests which included science and mathematics, though his main interests were history and theology, he was the driving force behind Venice's resistance to papal power, and was famous throughout Europe as the author of a critical *History of the Council of Trent*. He became friends too with his pupil Giovanfrancesco Sagredo, a cleric and nobleman who became a diplomat and was later to be immortalised by Galileo in his controversial *Dialogo*.

Galileo's public lectures covered the agreed syllabus, but he also supplemented his income by private lectures on military engineering, mechanics and, possibly, on astronomy. In addition, he began production of a mathematical instrument, an improved type of 'geometric and military compass' designed to help in a host of military calculations. This he had constructed by a skilled instrument-maker in a workshop attached to the house he now occupied; it brought in extra income, and was followed by other instruments. Galileo supplemented these by sales of a booklet explaining the use of the compass, while he also wrote several books on the subjects of his lectures for use by his students. In this way he was able to offset his debts caused by having to pay a dowry when his sister married, and support his household, for in Padua he had taken a mistress, Marina Gamba, who bore him two daughters and a son.

It is clear that by the time he had been at his new university for five years, Galileo was convinced in his own mind of the correctness of the Copernican theory of a Sun-centred universe, for when he received a copy of Kepler's *Mysterium cosmographicum* ('Cosmographic Mystery') in May 1597, he wrote expressing his support. However, when Kepler replied urging him to make his opinion public, Galileo allowed the correspondence to lapse. It seems that at this time Kepler, as an astronomer, was convinced of the truth of Copernicus' theory because of its elegant and straightforward explanation of the motions of the planets. However, Galileo's support was based solely on his mistaken belief that a rotating and orbiting Earth would together account for the behaviour of the tides; he had no other evidence, and so was unwilling to commit himself further. His main interests were still in mechanics. It is true, though, that in 1604 he became stimulated by the sudden appearance of a new star ('Kepler's Nova'), an event we now classify as a supernova and know to be due to the explosion of a star previously too dim to be visible to the eye. Galileo took an active part in the arguments which followed its appearance, siding as we might expect with those who said that the supernova disproved Aristotle's dictum that the heavens were changeless. Then four years later, an event occurred which was at last to cause Galileo to come out in open support of Copernicanism and, incidentally, to change his whole life. This was the announcement of the discovery of the telescope.

In October 1608 a Dutch spectacle-maker Hans Lipperhey, applied to his government for a patent to allow him to be the sole maker of instruments 'for seeing at a distance'. Sarpi soon got news of the invention from his diplomatic contacts and when Galileo was on a visit to Venice in July 1609, told him about it. Galileo, was somewhat sceptical, but Giacomo Badovere, an old student then in Paris, confirmed not only that such devices existed but also that they were on sale there. Next Galileo heard that an instrument had arrived in Padua and rushed back there, but the telescope and its owner had left. However, Galileo received a good enough description so that, with his previous knowledge of the way spectacle lenses aided vision, he was able to work out the optical effect of using two lenses together, and then construct a telescope in his own workshop. Like the telescopes then on sale, it was not a good one: it magnified only a little, gave a hazy field of view which was sharp only at the centre, and even then was bedevilled by coloured fringes. More in the nature of a toy than an optical instrument, it nevertheless demonstrated the basic principles. But with this behind him, Galileo then set about constructing optically improved versions, grinding his own lenses with immense care not given to those used in spectacles, and masking the front one in each instrument so that light could not pass through its outer regions. As a result he improved the image so that the whole field of view was now sharp and could bear greater magnification.

Meanwhile, the Doge of Venice was offered at an exorbitant price a telescope which magnified only three times. Knowing that Galileo's experiments were proving successful Sarpi, who had been asked his opinion, suggested the offer be declined. By August 1609 Galileo had a telescope that was good optically and magnified nine times. Of practical use to a maritime power like Venice, Galileo wisely presented it to the Doge. In return he received a life tenure of his chair at the university, and a quite unprecedented salary for a mathematician. However, when the official document setting out the new conditions arrived, Galileo found them to be different from those he had received verbally, and thereupon decided to press his application for a post at the Tuscan court back in his beloved Florence. This was to prove an unfortunate decision.

Meanwhile, back in Padua, Galileo managed to push the magnifying power up to thirty, and in January 1610 turned it to the heavens. The results were to make him famous. In the first place he saw that the Moon, far from being a perfect featureless body as Aristotle had claimed, was covered with craters and mountains. Indeed, he could even measure their heights by calculating the altitude of the Sun above the Moon's surface and determining the lengths of the shadows which the mountains cast. The Milky Way he found to be composed of separate stars – unknown to the Greeks, this clearly demonstrated that they were not aware of every scientific fact – while the planet Jupiter was accompanied by four orbiting satellites. This last put paid to the argument that the Earth could not move in space because the Moon would be left behind if it did. After all, in Galileo's opinion, what was true for Jupiter must, surely, be true for the Earth. These were momentous discoveries, and no time was lost in publishing them. They appeared in his *Sidereus Nuncius* ('The Starry Messenger') which came out early in March 1610. Later Galileo observed that the planet Venus displays phases, which only went to strengthen his belief that the observations confirmed the truth of the Copernican theory.

Though not usually a man of much tact, nevertheless Galileo was diplomat enough to name Jupiter's satellites the 'Medicean Stars' in honour of the Grand Duke Cosimo II de' Medici of Tuscany, for he had it in his mind to return to Florence if this proved possible. Soon after he was indeed invited to return there as philosopher and mathematician to the Grand Duke, and to be chief mathematician of the University of Padua. By the summer of 1610, and against the advice of his friends, he resigned his chair at Padua and returned to Tuscany, which was intellectually hidebound and not to be compared with Venice in matters of academic freedom.

To begin with all seemed to go well. Galileo visited Rome, had his discoveries confirmed by the Jesuits of the Roman College, was favoured and feasted. Yet soon after his return to Florence, controversy began with the pro-Aristotelians who were jealous of his position at Court. The argument was about the behaviour of floating bodies, but Galileo emerged triumphant, producing a book which, by argument and experiment, extended the work begun by Archimedes. In all this he was supported by Cardinal Maffeo Barberini, a mathematician later to become Pope.

Another controversy arose with the Jesuit, Christoph Scheiner, who had used a telescope to observe the Sun and had detected spots on the solar disc. In keeping with Aristotelian tradition that the Sun was an unblemished body, Scheiner believed these 'sunspots' to be tiny planets orbiting the Sun. Galileo not only opposed Scheiner's interpretation but also claimed that Scheiner had no priority in his observations; he, Galileo had seen sunspots a year earlier. Deep enmity followed this exchange.

Worse was to follow, for theological objections to the Copernican theory were raised during a Court dinner. Galileo was not present but wrote letters about the matter, taking the view that neither the Bible nor science could speak falsehoods. As Cardinal Baronius had recently put it: 'The Bible teaches the way to go to heaven, not the way the heavens go.' But there were those whose objections could not be silenced, and against the advice of friends and even of the Tuscan ambassador, Galileo set off again to Rome, this time to clear his own name and plead the Copernican cause. He was successful on his own account but not on Copernicanism, being instructed neither to hold nor defend the theory.

Wisely Galileo obeyed and threw himself into the less controversial world of physics. Then, three years later, in 1618, three bright comets appeared; these gave rise to many pamphlets of which one, by the Jesuit Orazio Grassi, raised Galileo's ire, and he attacked it. Though he used the name of a pupil rather than his own, Galileo's hand was clearly recognised in the pamphlet he produced. Grassi replied with a polemic with the result that Galileo wrote *Il Saggiatore* ('The Assayer'), a brilliant rebuttal, using all his powers of ridicule. One example of this biting wit is too good to miss. To keep up the myth of anonymity, Grassi had used the name Sarsi, and Galileo wrote:

> If Sarsi wants me to believe from Suidas that the Babylonians cooked eggs by whirling them rapidly in slings, I shall do so; but I must say that the cause of this effect is very far from that which he attributes to it. To discover the truth I shall reason thus: 'If we do not achieve an effect which others formerly achieved, it must be that in our operations we lack something which was the cause of this effect succeeding, and if we lack one single thing, then this alone can be the cause. Now we do not lack eggs, or slings, or sturdy fellows to whirl them; and still they do not cook, but rather they cool down faster if hot. And since nothing is lacking to us except being Babylonians, then being Babylonians is the cause of the eggs hardening.'[1]

[1] Translation from Stillman Drake and C. D. O'Malley, *The Controversy on the Comets of 1618*, Philadelphia, 1960.

With the book Galileo not only triumphed once again over his opponents, he also proved he could write something good to read. *The Assayer* was a great success; even the Pope had it read to him at meal-times and is said to have enjoyed it. But this was a new Pope, Urban VIII, Galileo's one-time friend Maffeo Barberini. Suddenly the future looked bright, for if a mathematician was now at the helm surely he would be amenable to rescinding the Church's strictures against Copernicanism. He of all men ought to acknowledge what was surely overwhelming evidence. Galileo therefore resolved to visit Rome yet again to plead his cause.

In April and May 1624 he was received with affection by the Pope on a number of occasions, but the outcome of their discussions was not all Galileo had hoped. Barberini had set his ecclesiastical authority above his feelings as a mathematician. In the event, therefore, Galileo had to be content with permission to publish his views but with the proviso that he reached no firm conclusion in favour of Copernicanism; he must remember, said the Pope, that God could create the universe any way he wished and cause it to display those characteristics that Galileo would insist on calling his proof of a moving Earth. With this he had to be content. Nevertheless, Galileo had made some headway, and he returned to Florence determined to take advantage of his permission to write about the Copernican theory.

Galileo did not treat this commission lightly; he worked on his text for the next six years. At last, in 1630, his book was complete, though an official licence was needed before it could be published. This took time, and only in February 1632 did it come out. Bearing the title *Dialogue on the Two Chief World Systems – Ptolemaic and Copernican*, it was soon to raise a storm of controversy.

Writing in the familiar form of a discussion, Galileo had chosen three protagonists – Simplicius, Salviati and Sagredo. Simplicius was named after a sixth-century pro-Aristotelian, and cast here as an unimaginative university academic. Salviati, pro-Copernican in his views, was modelled on Galileo's deceased friend Filippio Salviati, and Sagredo on another, the nobleman Giovanni Sagredo. In the discussions, Sagredo's task was to act like any leisured and broad-minded Italian intellectual, judging the arguments on their merits. The discussion was divided into four 'days', the first describing the Aristotelian case, the second and third the daily rotation of the Earth and its orbital motion round the Sun, while the last was concerned with Galileo's theory of the tides, which he still believed to be the crowning argument in favour of

Copernicanism. Yet true to the Pope's wish, the argument for divine omnipotence closed the book, though tactlessly Galileo put it into the mouth of Simplicius.

Written not in Latin, the *lingua franca* of the learned world, but in Italian so that it could reach everyone who could read, the *Dialogue* was a dangerous book. So dangerous that Galileo's enemies seized on the fact that it was Simplicius to whom the Pope's argument had been given, and persuaded Barberini that he had purposely been ridiculed. The upshot then was that Galileo was summoned to Rome by the Inquisition to answer charges that he had disobeyed the formal injunction of 1616 not to teach or hold the Copernican theory.

At this point Galileo was offered sanctuary back in Padua, for the Venetians were independently minded and powerful enough to resist Rome, but he would not go. Though he believed his Church was in error to pronounce on scientific matters, Galileo was far too devout a churchman to disobey its commands. In the end, therefore, he was put on trial, though it soon appeared that there were discrepancies in the evidence against him. Nevertheless, so powerful was the Inquisition and so deeply were they now involved, that Galileo was privately advised to plead guilty and throw himself on the mercy of the Court. This turned out to be wise advice and, shocked by the whole business, Galileo gratefully accepted it.

His sentence was nowhere near as stern as it might have been, but it was hard enough in practice. Galileo had to dress in the garb of a penitent and make a public recantation of Copernicanism; thereafter he had to submit to permanent house arrest. But at least he could return to his home at Arcetri, a few miles outside Florence. And in spite of rumours which circulated later, Galileo was neither tortured nor even shown the instruments of torture to persuade him to confess. Nor did he ever mutter 'Eppure si muove' ('Yet it still moves') at his recantation. He was not so foolish, and no such story was ever current in his lifetime; it was a later invention of the 1640s.

Besides sentencing Galileo, the ecclesiastical authorities banned all Galileo's books, including the *Dialogue*, but this was almost totally ignored. Merchants, nobles and even prelates vied with each other to buy copies on the black market, at prices rising anything up to twelve times the book's original value.

Galileo himself seems to have recovered remarkable quickly after his mental ordeal. He had other important work to do, and though now seventy years of age, set about it with vigour. Wise enough to cease promoting Copernican ideas, he turned back to his first love, mathematical physics. Five years later this bore fruit, when a manuscript on which he had been working was smuggled out of Florence and published in Holland. This was his *Discorsi e Dimostrazioni Mathematiche Intorno à Due Nuove Scienze* ('Discourses and Mathematical Demonstrations Concerning Two New Sciences'), which dealt with engineering science and pure mathematical physics. Though again written in the form of a discussion between Sagredo, Salviati and Simplicius, the book is not just a mathematical text; it also provides experimental confirmation of the mathematical results. Throughout it breaks new ground, and its discussion of motion is particularly notable. First of all, Galileo discarded Aristotle's division of motion into two types – 'natural motion' which controlled objects as they sought their natural place in the universe (the centre of the Earth in the case of heavy bodies) – and 'violent motion' which occurred when a body was forced to move in any other way (as, for example, when a horse pulled a cart). Instead Galileo brought all motion on Earth under one set of laws, which he analysed in mathematical detail, and then backed up with experiments. A few of these were 'thought experiments' – what would happen under certain specified conditions – but for the most part they were experiments which he had actually carried out himself.

This was a new and very powerful technique, which not only laid the foundations for all future work on the subject, but also brought Galileo very close to enunciating Newton's first law of motion, that a body continues in a state of rest or of uniform motion in a straight line unless acted on by outside forces.

Galileo's approach also allowed him to clear up an age-old problem, the motion of projectiles. According to Aristotle no body could undertake 'natural motion' and 'violent motion' at the same time. In consequence when a cannonball was fired, it would first move in a straight line due to the violent motion generated by firing the cannon, and then later suddenly change direction as natural motion took over, making the ball fall to the ground in a straight line. And though everyone knew that cannonballs did not behave in this way – they move in a curved path – Aristotle's teaching of the two separate motions was still illustrated in textbooks. Galileo, however, was able to demonstrate mathematically that not only would a body move in a curve – as observed – but also that the curve must be a parabola.

There is much else in this seminal book. The foundations are laid for the mathematical treatment of the strength of beams and structures, and in addition Galileo discusses a vacuum, the very existence of which Aristotle denied. Yet more than this, Galileo gives instructions both how to produce a vacuum in the first place – by constructing a cylinder with a close fitting plunger – and then how to measure its strength by hanging weights on the lower end of the plunger.

Rightly recognised as a masterpiece, the *Discorsi* laid the foundations for all future work in its new subjects, mathematical physics and mathematical study applied to engineering. Indeed, physicists and engineers have ever since remained in Galileo's debt for his pioneering approach to their problems, as valid today as it was almost 350 years ago. Gone at last was Aristotle's teleological approach to motion and to physics in general, his belief that everything could be explained by its natural purpose within the scheme of the universe as he understood it. Now it was replaced by mathematical analysis and experiment going hand in hand, an immensely powerful and timeless approach, independent of whatever ideas about the universe the scientist might propose. In achieving this Galileo can be said to have broadened and consolidated Kepler's mathematical approach to astronomy, as well as paving the way for Newton and his successors in the physical sciences.

The *Discorsi* was smuggled to Holland with the connivance of the French ambassador, the Duke of Noailles, and in his preface Galileo claimed that publication was arranged without his knowledge. All the same he dedicated the book to Noailles, and no-one was really deceived, including the ecclesiastical authorities; but honour was satisfied. And as far as Galileo was concerned he had conquered his conquerors.

In 1636 and 1638 Galileo received visits from the philosopher Thomas Hobbes and from John Milton. Milton was interested in Galileo's power to publish, even under restriction, but Hobbes was more impressed by Galileo's science; he was, Hobbes claimed, '. . . the first that opened to us the gate of natural philosophy universal . . .'

Even before the manuscript of the *Discorsi* was smuggled out of Italy, Galileo began to suffer a severe eye infection, and by June 1637 he was blind in his left eye. By December he was totally blind. Tended by his students Vincenzo Viviani and Evengelista Torricelli, he managed to keep up his correspondence, and his mind remained active; he dictated the details of a pendulum clock mechanism to his son, Vincenzo, and devised a calculator for determining the future positions of Jupiter's satellites. But these were his last achievements. Early on 8 January 1642 he died.

In spite of his readiness to forgive, Galileo's quick temper and caustic tongue made him enemies, though it is clear that he was loved by his students and by a host of friends and acquaintances. As to his quarrel with the Church, there is something to be said for the view that Galileo was a martyr of science, though really that is only part of the story. More correctly he was a martyr to entrenched opinion. The blinkered Aristotelian academics were as much his enemies as narrow minded churchmen. Neither could understand the independence of thought necessary to formulate a scientific picture of the world. Galileo was, of course, too penetrating an intellect not to appreciate this, and we can safely leave the last word to him:

> In the matter of introducing novelties. And who can doubt that it will lead to the worst disorders when minds created free by God are compelled to submit slavishly to an outside will? When we are told to deny our senses and subject them to the whim of others? When people of whatsoever competence are made judges over experts and granted authority to treat them as they please? These are the novelties which are apt to bring about the ruin of commonwealths and the subversion of the state.[2]

[2] From G. de Santillana's translation of Galileo's *Dialogue on the Great World Systems*, University of Chicago Press, Chicago, 1953, p. ix.

JOHANNES KEPLER:
The New Astronomy
Jim Bennett

The progress of science involves two processes; theoretical innovation and acceptance of those innovations. The latter often trails the former. Such was the case with the astronomical ideas of Copernicus who, in 1543, argued that the Sun lay at the centre of the planetary system. For the rest of the sixteenth century his theory had few supporters.

Johannes Kepler (1571–1630) was born in Weil der Stadt and studied at the Protestant University of Tübingen, intending to enter the Lutheran clergy. Before completing his studies, however, he was sent as a teacher to Graz in Austria where he took up his career as mathematician and astronomer. In 1600 he moved to Prague where he wrote his revolutionary *Astronomia Nova* (New Astronomy) in which he placed the heliocentric (Sun-centred) system on a sound mathematical footing and began the task of reuniting the physical and computational sides of astronomy.

When a scientist's name becomes attached to laws commonly taught in school physics, celebrity is assured among a public many times more extensive than the specialists in the history of science. But Kepler's three laws of planetary motion have been a mixed blessing. Planets move in elliptical orbits with the Sun at one focus; the line joining each planet to the Sun sweeps out equal areas in equal times; the squares of the planets' periods are proportional to the cubes of their mean distances from the Sun. These laws have served to make Kepler's name widely known, but have given a distorted picture of his work – unhistorical and out of sympathy with his true goals. Kepler himself had no concept of 'three laws'; their selection from among his writings has given them a significance he could not have recognised, and that could be seen only when the planetary theory of Newton was published more than fifty years after Kepler's death. His contribution to shaping the discipline of astronomy, on the other hand, has an importance that transcends the many relationships – the 'three laws' included – he discovered among the planetary motions.

To appreciate this we must begin by understanding the situation of astronomy as Kepler found it. Briefly there were two distinct approaches to the heavens, and those who pursued one or the other recognised a large measure of separation between them. On the one hand there were the mathematical astronomers, who dealt with geometrical descriptions of planetary motion, with predictions of planetary positions and with applications in such areas as calendar construction, timekeeping and navigation. On the other hand there were the 'natural philosophers', who dealt with such questions as the physical nature of the heavens, the mechanisms of its motion, and its general organisation in qualitative terms and on a cosmological scale. We shall call these two disciplines mathematical astronomy and physical cosmology.

To discover the source of this division we must go back to the ideas of the ancient Greek philosophers Plato and Aristotle. Plato taught that the heavens were constructed according to a perfect geometrical model. This model was fully realised only in a transcendental eternal realm, which was experienced by man's immortal soul; it could never be perfectly expressed in the physical world, which by its nature must fall short of the true model. Thus for Plato the true astronomer was a pure mathematician – a geometer. He dealt with the perfect model, not with its imperfect, physical manifestation. Sense experience of the physical world provided clues to the original model, but it was truly accessible only through the intellect, which Plato regarded as a 'memory' of the soul's experience of the transcendental realm. The mathematical astronomer was therefore unmoved by questions of material construction or physical cause.

Aristotle had little time for all this, and took a more common-sense view of things. For him the physical reality of sense experience was all we could know and all there was to know. Accounts of the heavens that reflected this reality must therefore be expressed in physical terms. Further, he held that mathematics did not get to the heart of things, that the world was conceived and constructed in physical rather than mathematical terms. A proper explanation would therefore not be mathematical but qualitative – constructed in terms of causes, which for Aristotle involved what things were made of, as well as what invoked their characteristic behaviours or motions.

It was from these different understandings of the world that the distinct disciplines of physical cosmology and mathematical astronomy were derived. A further assumption of the mathematical astronomers will be important for Kepler. We have seen that Plato's model was to be perfect, so the motions it comprised must be the most beautiful and harmonious motions possible. According to Plato this meant that only motion that was uniform or regular and circular could be admitted to the perfect model. The problem with this was that the observed motions of the planets – even allowing for the imperfections of the physical world – could not readily be accommodated by motions that are uniform and circular. Among various forms of anomalous or deviant behaviour, which challenged any modelling in Platonic terms, the most obvious was planetary 'retrograde' motion – the habit of stopping periodically in an observed path and moving 'backwards' for a time, before resuming the customary eastward motion through the fixed stars.

The outcome of centuries of this modelling exercise was a mathematical astronomy of enormous complexity, based entirely upon circles – rotating circles whose centres were themselves rotating upon larger circles – built into a grand system that could predict planetary observations with reasonable precision. It was not, of course, open to physical questioning: to ask for a physical account of the system of interdependent circles was to misunderstand the discipline of mathematical astronomy. Such questions had to be directed at the cosmologist, and cosmological theory had settled for a system of rotating spheres – physical, transparent spheres, centred on the Earth, and transmitting motion mechanically from the outer spheres to the inner ones. Each planet was attached to a sphere and carried round by its motion, and close to the outer limit was the sphere of the fixed stars.

Now there were very obvious discrepancies between the accounts offered by the two disciplines. They agreed on some general suppositions: that the universe – the 'cosmos' – was spherical, that its motions were circular, that the Earth was stationary at its centre. But beyond this they diverged. How, for example, could *rigid* spheres be compatible with Ptolemy's circles (epicycles) rotating on circles (deferents), which must vary a planet's distance from Earth. One way of handling such discrepancies was to say that, not only did mathematical astronomers not deal with physical questions, but they were not primarily concerned with questions of truth. Their major concern was predictive accuracy, and the model itself need not be true so long as it gave the right answers. While not everyone subscribed to this idea, it had been encouraged by a geometrical device introduced by Ptolemy, one that Kepler himself was to adopt for a time.

Ptolemy had discovered that he could improve the predictive accuracy of his geometry by a trick that transgressed the canon of Platonic astronomy. When a planet's motion is uniform in a circle, it appears uniform only from the circle's centre – that is, angular velocity is uniform about the centre, or the line drawn from the centre to the planet (the radius vector) moves through equal angles in equal times. Ptolemy found he could achieve better results by identifying a point (known as the 'equant point') displaced from the centre, and postulating equal angular velocity about the equant point. This meant, of course, that the planet's motion (or, to be precise, the motion of the centre of the planetary epicycle) was not uniform on the circle. One of the primary Platonic rules had been ignored – evidence, surely, that mathematical astronomers were concerned only with results, that they were pragmatists.

Not all mathematical astronomers were happy with the role of pragmatist, and some tried to remove anomalies in the geometrical model. Nicolas Copernicus, for example, was particularly unhappy with the equant device, and managed to replace the equant point for each planet by a further epicycle. He is better known, of course, for making the Earth a planet – for setting the Sun in the centre of his planetary model, with the Earth in motion around it. The full account of his theory was published in his *De Revolutionibus* of 1543, and it was to have a profound influence on the young Kepler.

Whatever their views on pragmatism in mathematical astronomy, practitioners in both disciplines subscribed to the separation between it and physical cosmology. Copernicus, for example, was putting forward a radically different geometry of the heavens – one which breached even those very general assumptions held in common with Aristotelian cosmology. He was therefore obliged to substitute a cosmology of his own, and this he did in the first of the six books of *De Revolutionibus*. It was based on a system of spheres, altered as little as possible from the cosmology of Aristotle, and was a qualitative account, showing as before only the minimal congruence with the detailed mathematical astronomy that followed in the bulk of the work. Even in the radical proposal of Copernicus, the separation of disciplines was maintained, and the conventions that governed the nature of astronomy were preserved.

Kepler's achievement is so bound up with breaking this mould that we can only now begin to consider it. He was born in Weil der Stadt near Stuttgart in 1571, and brought up as a Lutheran. His father was at times a mercenary soldier – 'quarrelsome', Kepler said, and 'criminally inclined' – who abandoned his family in 1588. He had more connection with his mother, but described her as 'thin, garrulous, and bad-tempered'. He went to a number of schools, before matriculating at the University of Tübingen in 1587, with a scholarship supported by the Duke of Wurttemberg.

It was at Tübingen that Kepler began to study astronomy, under the influence of the professor of astronomy, Michael Maestlin. For some time he saw his true vocation in the ministry, and studied theology after receiving his master's degree in 1591. However, he was chosen as mathematics teacher in the Lutheran school of the Austrian town of Graz, and in 1596 he published his first book, the *Cosmographical Mystery*. It was a remarkably arrogant work for a man of twenty-five and the subject was Copernican astronomy.

It must be emphasised that whereas a number of mathematical astronomers of the late sixteenth century were prepared to use Copernicus's geometrical system for calculating planetary positions and tables, those who accepted the reality of the Sun-centred cosmos were very few. The arguments in favour rested principally on aesthetic considerations, based on the greater harmony and rationality of the system. Planetary retrograde motions, for example, were very neatly explained by Copernicus as an illusion created by observing the other planets from a moving platform. On the other hand, the absurdity of a moving Earth was only the most obvious of many difficulties. Maestlin was one of the few realist Copernicans, and Kepler had been readily won over. Kepler later described his 'conversion' to Copernicanism in fairly typical language, writing that he 'attested it as true in my deepest soul, and contemplated its beauty with incredible and ravishing delight'.

By the time he wrote the *Cosmographical Mystery* he was asserting that, whereas Copernicus had argued on the basis of observations, he, Kepler, was able to demonstrate the truth of the system from first principles – to reveal the geometrical rationale behind its construction and to show, as he put it, why the number, sizes and motions of the orbs or spheres 'are as they are, and not otherwise'. Why are there only six orbs (one carrying each planet), why are they arranged and moving as we find them?

Kepler's explanation for the number and sizes of the orbs, announced in the *Cosmographical Mystery*, is ingenious. It is a fact of geometry, known from antiquity, that there are only five regular or perfect solids all of whose sides are the same regular figure.

55

These are the tetrahedron (composed of four equilateral triangles), the cube (six squares), the octahedron (eight equilateral triangles), the dodecahedron (12 pentagons) and the icosahedron (20 equilateral triangles). By using these solids and the most perfect figure of all, the sphere, Kepler accounted for the planetary arrangement as follows.

Within the sphere for Saturn inscribe a cube, so that its eight vertices touch the sphere. The sphere for Jupiter is found to inscribe this cube, so as to touch its six planes. The tetrahedron is inscribed within the sphere for Jupiter, and is found to circumscribe the sphere for Mars. The dodecahedron, the sphere for Earth, the icosahedron, the sphere for Venus, the octahedron and the sphere for Mercury are arranged in turn, and the model is complete. Its appeal is obvious. There are only five regular solids, there are five spaces between the planets, the solids fit the spaces. 'There', says Kepler triumphantly, 'you have the reason for the number of the planets.'

It is important to introduce a central element in Kepler's astronomical thought, which might be called its 'devotional' aspect. For Kepler the natural world was a key to the character of God. He said that the three motionless elements in the Copernican cosmos – the Sun, the sphere of the fixed stars, and the space between – were an image of God's triune nature. He believed that the arrangement of moving elements – the planets – would also reveal something of the Divine, and in their perfect geometrical model God is revealed as the supreme geometer. The devotional impetus for Kepler's quest for rationality in the cosmos is a powerful motivation underlying all his work, and it helps to account for some of his characteristic attitudes.

If the number and sizes of the spheres had been explained, their motions would be more difficult. Kepler does not take this problem very far in the *Cosmographical Mystery*, but he does introduce the idea that the Sun is somehow the cause of planetary motion. The Danish astronomer Tycho Brahe, who had built a fabulous observatory on the island of Hven in the Danish Sound, and amassed a unique observational record with his array of large astronomical instruments for measuring angular distances with unprecedented precision, had demonstrated that comets move in the region of the planets. For Aristotle, comets had been meteorological phenomena, close to Earth, but as Tycho observed, if comets moved without apparent hindrance among the planets, there could be no such thing as solid, material spheres. Now Copernicus had shown that planets farther from the Sun move more slowly than those close by, so that Kepler suggests that their motion results from a solar influence. Since it acts at a distance, he thinks of this influence in animistic terms; he calls it an 'anima motrix', a motive soul.

Kepler naturally sent copies of his work to eminent mathematicians and astronomers. He sent one to Galileo, who replied to say that he too was a Copernican – but not yet in public. Tycho wrote Kepler an encouraging reply, with an invitation to visit him, though in truth he disliked Kepler's *a priori* way of doing astronomy.

Two factors together brought about a profound change in Kepler's situation before his next astronomical work was published. First, the religious and political influence of the Counter-Reformation made his position untenable in Graz. In 1598 most of the Protestant teachers were obliged to leave Catholic Graz, and Kepler, having refused to change his faith before a commission on 2 August 1600, was banished from the town. Second, the model of the perfect solids did not, in fact, fit the cosmos as well as Kepler would have liked. He was sure that this was due simply to the inaccuracy of his data, and was anxious to gain access to the observational records of Tycho. Since no position was available at Tübingen, Kepler saw his only future in the service of Tycho, and in desperation he set out with his family before Tycho's acceptance had been confirmed. Tycho himself was now installed in Prague as Imperial Mathematician, having lost his patronage by the Danish crown. Kepler was later to attribute this move of Tycho's to the work of Providence, in bringing Tycho and his observations within reach.

Kepler was assigned the task of calculating the orbit of Mars, which had defeated Tycho's previous assistant, Longomontanus. But in the very year of Kepler's arrival, Tycho died, and Kepler's access to the enormous collection of observations was unrestricted. The conclusions to his work on Mars were published in 1609 in a classic astronomical work that challenged the very foundations of the discipline. This challenge is issued in the title itself: *New Astronomy, Based on Causes, or Celestial Physics, Expounded in Commentaries on the Motion of the Planet Mars*. Kepler was to raise physical questions as part and parcel of astronomy, to demand that a full astronomical explanation was not only a geometrical model, but also a physical account of how it worked.

An important change from the *Cosmographical Mystery* was that the influence that moved the planets was no longer a soul, but a physical force – a 'vis motrix' (motive force) was substituted for the 'anima motrix' – and this force was magnetism. The idea had come from the English natural philosopher William Gilbert, who had published his famous account of magnetism, *On the Magnet*, in 1600. As well as experimenting with magnets, Gilbert speculated on the possible cosmic role for magnetism, having discovered that the earth herself had the properties of a huge magnet. It was this cosmic role for magnetism that Kepler developed in his *New Astronomy*.

Kepler's first move in trying to handle the mathematics of the Martian orbit was to re-introduce the equant point that Copernicus had banished, at this stage combined with a circular path. With Kepler the Platonic canon was going for good. He preferred to recognise that the planet's motion was non-uniform, by using an equant, rather than to 'disguise' the fact, as Copernicus had done, with an epicycle.

The equant and circular path gave good agreement with Tycho's observations of the longitude of Mars. The distance calculations, however, which involved the latitude observations, were less satisfactory, and Kepler was scrupulous to an unprecedented degree. He regarded Tycho's observations as a gift from God – part of the providential design for his destiny; he knew how accurate they were, and he had to be honest to them. Here the devotional element in Kepler's astronomy was strong. To discover the divine geometry, revealed in the created universe, we must begin by accepting what God is showing us. We must first discover the true planetary paths, without trying to make them conform to what we think they should be. This left no room for the Platonic canon. Further, we must use the observational evidence honestly, carefully, humbly – in a devotional spirit: '. . . it is right', wrote Kepler, 'that with grateful hearts we should both acknowledge and put to use this gift of God. . . . These eight minutes alone [the latitude error] have pointed the way to the reformation of the whole of astronomy.' Here Kepler foreshadowed his introduction of non-circular planetary orbits.

All of this represented a very considerable change from the earlier *a priori* emphasis of the *Cosmographical Mystery*, a change occasioned no doubt by the impact of Tycho's observations. It was not, as we shall see, a final renunciation of the grand vision. Only Kepler's devotional approach to astronomy explains the apparent contradiction of scrupulous empirical care, rightly using the gift of Tycho's work, with the aim of discovering the perfect model of the supreme geometer.

Kepler next discovered that there was, in fact, no equant point;

there was no point from which Mars would appear to move with uniform motion. Having exhausted mathematical possibilities, he turned to physics. It was typical of Kepler's method in the *New Astronomy* that he tried to juggle with geometry and physics together, to move from one to the other in turn, and not to accept a conclusion from one side until he had a parallel argument from the other. The equant had failed to account for orbital velocity, but if the Sun is the physical cause of planetary motion, the force moving the planets might be supposed to decrease with distance from the Sun, and their speed (according to Aristotelian mechanics) will be proportional to this force.

On physical grounds Kepler now had an expression for orbital speed – it was inversely proportional to distance from the Sun – and he had a physical mechanism for it in cosmic magnetism. The magnetic influence of the Sun swept the planets around the solar system, and decreased with distance; Kepler was delighted when Galileo subsequently announced, from telescopic observations of sunspots, that the Sun really was rotating.

It was going to be difficult to apply this expression (the inverse distance law) to finding the time taken to traverse a portion of the orbit, since this would have involved summing an infinite number of radius vectors (remember there was no infinitesimal calculus). Kepler introduced a convenient approximation, by saying that the area enclosed by these radius vectors was an approximate measure of their sum, and therefore of the orbital time. The planet will, in other words, sweep out equal areas in equal times – an expression which is not mathematically equivalent to the inverse distance rule, and which Kepler regarded as only a mathematical approximation to the real physical situation. So much for 'Kepler's second law' – the name bears no relation to its role and status in the *New Astronomy*.

Kepler now had two expressions for planetary speed, but as yet no path. He was clear, however, that the observations would not allow a circle, and he played the old game of introducing an epicycle, combined now with the area law. These could generate a number of different oval shapes, and he eventually realised that a perfect ellipse, and the area law, would account for the observations. At this stage, however, this was only a mathematical result, and could not be accepted without a physical derivation.

In particular, the epicycle was a mere mathematical device and made no physical sense. When Kepler turned once again to consider the physical situation, he found it more satisfactory to think of the planet moving along the radius vector to the Sun – now

approaching, now receding, in a 'libration' caused by the interaction of the planetary magnetic axis and the magnetic Sun.
Gilbert had, after all, demonstrated that the planet Earth was a
magnet. The detailed physical argument, superimposing the libration caused by Mars's magnetic axis on the magnetic effect of
the Sun's rotation, concluded with an elliptical path.

Since geometry and physics now both yielded an ellipse, Kepler
was prepared to accept this as the true path: 'With reasoning
derived from physical principles agreeing with experience there is
no figure left for the orbit of the planet except a perfect ellipse.'
This is what we call his first law.

The laws are important parts of the detailed structure of
Kepler's astronomy, but its real achievements lie at a more fundamental level. We can take two examples. First there is the
concept of an orbit, which hardly existed before Kepler's mathematical astronomy. Attention had always been focused on the
geometrical components – the epicycles and deferents, or what
Kepler called 'the useless furniture of fictitious circles and
spheres'. When challenged that he had transgressed the canon by
suppressing circular motion, Kepler made precisely this point –
that his resultant paths were not very different from those of
Copernicus, and the difference lay rather in the object of study:
'When you speak of the components of motion, you speak of
something which is only imagination, and which does not exist in
reality; for nothing performs the circuits in the sky except the body
of the planet itself.'

Further, this was linked to his devotional attitude to astronomy.
If he was trying to discover God's plan in creation, there was little
point in conditioning the result beforehand by human inventions.
'The simplicity of nature', he wrote, 'must not be judged by our
imagination.' The divine architecture was not to be restricted by
human metaphysics.

Second, Kepler insisted on the unity of physical cosmology and
mathematical astronomy. The interplay of physical and geometrical arguments in the *New Astronomy* contravened all the conventions, and caused bewilderment, outrage and neglect. Even
Kepler's old Copernican teacher Maestlin reminded him that
'Astronomical questions should be treated astronomically, by
means of astronomical rather than physical causes and
hypotheses. Astronomical calculation is based upon geometry and
arithmetic, not on physical conjectures, which disturb the reader
rather than informing him.'

What of the third law? It appeared in Kepler's final work, *The*

Harmonies of the World, published in 1619. (The more popular *Epitome of Copernican Astronomy* appeared in five books between 1617 and 1621.) Kepler was once more caught up in the search for the divine aesthetic expressed in creation, now based on the empirical results derived from Tycho's observations. His original aim, remember, had been the reason for the number, sizes and motions of the orbs being 'as they are, and not otherwise'. Number and sizes were still explained by the regular solids, adapted now to elliptical orbits. This model never lost its appeal for Kepler; it was the design used by God for the basic framework of the cosmos, to be filled out in detail by other harmonies and relations.

While the motions had been described in the *New Astronomy*, and given a physical explanation, they did not yet have a geometrical rationale; the results of Kepler's work on Mars had yet to be applied to his ultimate goal of uncovering the architecture of creation. Kepler's solution derived from an old Pythagorean idea; he was convinced that planetary motions somehow exhibit harmonic ratios that mirror the harmonies of music. He eventually related the musical 'note' of a planet to its angular velocity with respect to the Sun and found that the ratios of maximum angular velocity, when the planet was closest to the Sun, and minimum angular velocity, when farthest away, in all cases corresponded with consonant musical intervals. He went on to draw in musical notation the 'tunes' played by each planet, and to work out combined 'harmonies' – the notes that all six planets play in certain configurations.

Other relations drew aspects of the whole system together, and linked them in an interdependent structure, so that nothing was left to chance. The overall periods of the planets, for example, should bear some relation to their distances. Kepler eventually decided that the squares of the planets' periods were proportional to the cubes of their mean distances, and we – with staggering disregard for its context in the *Harmonies* – have the nerve to call this 'Kepler's third law'. Like the other two, it can be singled out in this way only with hindsight; for Kepler its significance depended on a vastly ambitious programme of devotional astronomy.

We said that in the *New Astronomy* the Platonic canon was dead, and this is true at the level of detailed rules and conventions. In the *Harmonies* we find a powerful expression of the Platonic vision. The world was created according to a perfect geometrical model – informed by a divine architecture and regulated by musical harmonies. Kepler's devotional astronomy compelled empirical accuracy and humble acceptance of the record of creation, unencumbered by human metaphysics, but its ultimate aim was an aesthetic – almost mystical – appreciation of the divine plan. This was possible, said Kepler, because we are made in the image of God, and our geometry, as Plato himself had known, was not learnt of this world, but was a portion of the divine image. It mattered not, wrote Kepler, if the *Harmonies* lay unread for a hundred years; God had been waiting for man's understanding since creation itself.

If there were few committed Copernicans during Kepler's early career, fewer still were won over to the full Keplerian vision of the cosmos. Nonetheless, Newton inherited not only the empirical relationships Kepler had wrested from Tycho's observations – we call them the three laws – but also a new programme for astronomy. The mathematical forms of the uncompounded planetary orbits were to be derived, on physical grounds, from the first principles of celestial dynamics. Whatever Newton's disclaimers regarding the physical status of his gravitational theory, everyone else understood that this was the programme of the *Principia*.

Antoine-Laurent Lavoisier
(1743–1794)

Isaac Newton (1642–1727)

Joseph Priestley (1733–1804)

William Harvey (1578–1657)

WILLIAM HARVEY:
The Discovery of the Circulation of
the Blood
Andrew Cunningham

Medicine was put on a rational scientific footing by the Greeks; Hippocrates was a great pioneer of clinical medicine and Galen left a body of theory which doctors imbibed until the Renaissance and beyond. But the power of medicine to overcome disease remained strictly limited. The agents of much disease – micro-organisms – were invisible to doctors without the microscope, and ignorance of human anatomy and physiology was widespread.

William Harvey (1578–1657) worked to overcome this ignorance through exploration of physiological processes. He developed a radically new theory for the circulation of the blood and made profound discoveries in the science of embryology. But while Harvey's techniques served as fine examples of systematic experimentation, his discovery of the circulatory system made no contribution, in the short-term, to the fight against disease.

William Harvey (1578–1657) discovered the circulation of the blood in about 1618 or so, and he published on it in 1628. The discovery surprised him. For neither he nor anyone else had ever suspected that the blood circulated round the body in a full circuit, and neither he nor anyone else had ever set out to find evidence for it doing so. But if the discovery was a surprise for him, it was not exactly an accident, for Harvey had been engaged in a private research programme on animals since the very early 1600s, and continued such work until his death. The interesting question therefore is what was Harvey doing – what was his research – in the course of which he unexpectedly discovered the circulation? We must remember that the fact that the blood circulates is not in any way obvious. Even today we cannot actually observe the blood circulating: at best, under artificial laboratory conditions, we can see the movement of some blood in one direction through some vessels of a living creature, but not the movement of all the blood continuously round the body, pumped by the heart out through the arteries and coming back through the veins. That the blood

65

circulates is (and has to be) a conclusion come to by inference from certain phenomena: and if we want to convince other people of it, we need to put forward an argument. This was Harvey's position too. The argument he put forward to persuade his readers had to be the more careful because all his contemporaries believed that the blood moved in quite different ways from that which he was suggesting.

To discover the nature of Harvey's anatomical project and why he took it up, we need to look briefly at his early life. Harvey was born in Folkestone, and went to the King's School, Canterbury. From here he went to Gonville and Caius College, Cambridge, where from 1593 to 1597 he studied the normal arts and philosophy course. The curriculum was built on and around the works of the ancient Greek philosopher Aristotle (fourth-century BC). Harvey could have stayed at Cambridge to study medicine, but it was not the best place to do so. He chose instead to go to Padua in Italy, thus following a tradition of his college, and possibly helped by a college scholarship to do so. Padua was then the most important university in the world for medicine, with students coming to it from all over Europe. Padua was used by Venice as its university and run on business lines. When Harvey arrived there the university had just been rebuilt very handsomely, and included in the new buildings was a large anatomical theatre. Hitherto anatomical theatres had been built and dismantled every season, but this was the first permanent one in the world (and it still survives). Galileo was teaching at Padua while Harvey was there, but seems to have made no impression on the young student. But one man there, certainly, made an impression on him which lasted Harvey's whole life: Girolamo Fabrizi d'Aquapendente (for short we shall call him by his Latin name, Fabricius) who was the Professor of Anatomy from 1565 to 1613.

The medicine that Harvey learnt at Padua was perfectly conventional for the time: it was the medical system first created by Galen, a Greek physician of the second century AD who lived and worked in Rome, and whose works survive in remarkable quantity. Galen had had a very close knowledge of anatomy, and built his medical system upon it. As a result of the work of anatomists in the Italian Renaissance, especially that of Vesalius at Padua in the 1530s, demonstrations of human anatomy had become central to medical education. So, over a period of about two weeks every year, the bodies of one or more condemned criminals were dissected in front of the medical students at Padua in the grand new anatomical theatre. For the students this was a very exciting

66

occasion, and it provided them with an opportunity to see the parts normally hidden inside the human body with which their practice, as physicians, would later deal; they were seeing the human body as Galen had described it.

But what Fabricius the Professor of Anatomy was teaching (when he could be persuaded to lecture at all) was something different from this. For what Fabricius taught were the results of his own research, and this research was conducted not on the human body, and not with respect to improving medicine: it was thus not 'Galenic'. Instead, Fabricius' research was modelled on the works of *Aristotle*: Fabricius saw himself as reviving and practising anew an anatomical research programme that Aristotle had been conducting in antiquity. In reviving the most ancient research programme he could find (and one centuries older than that of Galen), Fabricius was doing a typically 'Renaissance' thing. This research took as its subject not 'man' but 'the animal'. Topics typical of this research programme were: the generation of animals, the local motion of animals (i.e., their locomotion), respiration. Fabricius published a treatise on each of these themes, just as Aristotle had done. In every case, Fabricius was investigating his topic with respect to 'the animal': hence he did his research on as many different kinds of animal as he could. The answers that he arrived at were applicable to *all* animals. If we take his work on respiration as an example, it is about respiration as a phenomenon of 'the animal', what its function is (why all animals respire, what role in the life of 'the animal' respiration plays), and the different means by which it is actually accomplished in different creatures. It is not 'comparative anatomy'; Fabricius' questions and answers are about 'the animal', its organs and functions, not about particular animals. Thus his work is about man, but only insofar as man is one kind of animal: man is not the centre of his attention. The point of this programme was to go further in the direction that Aristotle had marked out, to find out more: it was not intended to find fault with and replace the opinions of Aristotle or other 'ancients'. Young William Harvey was greatly excited by this programme, as we shall see: it is this research programme that he was to take up and practise for fifty years.

Harvey returned to England in about 1602 with his Paduan MD diploma. He set up in medical practice in London, building a clientele from those rich enough to afford a doctor's fees. He became a member of the College of Physicians of London; they appointed him their Lumleian lecturer on anatomy in 1615, a post he held for twenty-eight years. In addition to his private practice he also sought appointment as Physician to St Bartholomew's Hospital, a post he held from 1609 for thirty-five years. Here, one day a week he came and gave his advice to the poor for a small salary. From at least 1618 he was a physician to James I, and this was renewed under Charles I. This position was very dear to him, for he was a strong and loyal supporter of the royal cause. His loyalties were seriously tested, for during his lifetime occurred the Civil War. But throughout his life Harvey's commitment to the royal cause did not waver: he even appeared on the battlefield at Edgehill (1642) in his capacity as physician to the king. In his politics Harvey was, thus, conservative.

In the midst of this very busy life as a practising doctor, Harvey did something very unusual: he found time and space to undertake an enormous programme of research at his own expense. For fifty years, mostly at home, and with no institutional support, Harvey passionately and doggedly pursued anatomical research of the kind he had learnt from Fabricius: and when he began he was probably the only person in England to be doing anatomical research at all. One can see from this how impressed he must have been by that research programme. It was the Aristotelian programme of research into 'the animal', the only anatomical research tradition to which he had been fully exposed.

His teacher, Fabricius, had been the first person to revive, in its fullness, the Aristotelian form of research into 'the animal'; Harvey was the second. This is what made his work unique. For he thus had a different set of questions and a different viewpoint from all his contemporary anatomists. And it is this which was to lead him, unexpectedly, to discover the circulation of the blood. But the topics on which he began research in the early 1600s were not concerned with the blood, let alone its movement. They were the same as or similar to those of Aristotle and Fabricius: the generation of animals; respiration; the local motion of animals. One of his topics, however, was on the central organ of the body. It was, in Harvey's own words, on 'the motion, pulse, action, use and utilities of the heart and arteries'. Why should Harvey have chosen such a project? He later wrote that he was actually willing to publish on this topic because when Fabricius 'learnedly and accurately delineated in a particular treatise each part almost of animals, he left only the heart unattempted'; yet his reason for choosing it in the first place lay in the fact that it was a typically Aristotelian object of inquiry: according to Aristotle's characterisation of the animal body, the heart was the most important organ. As Harvey later put it, 'the heart of animals is the foundation of

68

life, the ruler of each of them, of the microcosm it is the sun from which all *vegetatio* derives, all vigour and strength emanates'. '*Vegetatio*' means the growth, maintenance and reproduction of the animal (i.e. everything except sensation, motion and thought): it refers to all those aspects of life which animals have in common with plants. Indeed the true object of all Harvey's anatomical inquiries here, as it had been of Aristotle's and Fabricius' before him, is the agency controlling this aspect of the life of the animal: what they all referred to as 'the vegetative soul'. In this sense the heart is the centre, source, principle, and origin of the life of the animal, and that is why Harvey set out to study it and its vessels.

When he began his research Harvey was wearing, as it were, two hats: he was an Aristotelian anatomist, and he was a Galenic physician. Under his physician's hat what would he have taken for granted about the heart and the movement of the blood? He would have believed that the veins constitute one system of vessels, based on the liver and distributing to the body the blood made by the liver from digested food: the veins thus constitute the *nutritive* system. The arteries, by contrast, contain blood which keeps man alive in the sense of being 'vivified': based on the (left side of the) heart, the arteries carry blood which has been impregnated in some way with the vivifying principle ('spirit' or *pneuma*) which the lung gets from the air. In other words, the arteries are part of the *respiratory* system, their pulsation being evidence of their role in the maintenance of life; and hence the Greeks had named the *arteria* after the air, *aer*, it contains. The visible and tangible differences between the two kinds of blood vessel would confirm all this: only the arteries pulsate, the walls of the arteries are thicker than those of the veins, the blood in the arteries is a different colour from that in the veins, and so on; and the fact that both systems reach all parts of the body is obviously because nutrition and vivification are necessary for all parts.

It was also known that the heart (as a whole) alternately expands and contracts. The expansion was known as *diastole* and the contraction as *systole*. It was assumed that when the beating heart *looks* larger, it is expanding, and vice versa; it was also assumed that expansion was the active stroke: that the chambers of the heart actively expand, and the blood rushes into them primarily to prevent a vacuum. Such movements were also known to be typical of the arteries. But the expansion and contraction of the arteries were thought of as movements independent of those of the heart: the arteries expand not because they are filled by blood being forced into them (as we now think), but because their own active

69

expansion *sucked in* blood from the heart. Thus the pulsation of the arteries was seen as an active movement of the arteries, and as taking place *alternately* with the beat of the heart.

In this arrangement the heart plays what, to us today, looks like a curiously anomalous role. For the *left* side of the heart is concerned with the respiration of the whole body: but the *right* side has the role simply of supplying nutritive blood to a single organ, the lung. In between the right and left chambers of the heart there is a wall (*septum*), keeping the blood of the two systems apart; but since the arterial system needs to be kept 'topped up' with blood, a small amount of blood seeps through this wall via some tiny pores. In the 1550s an anatomist in Rome, Realdo Columbo, had argued (for some very particular reasons) that the blood from the right side of the heart goes to the lung, becomes changed there, and then comes to the left side of the heart and thence enters the arteries; Columbo was not concerned to discuss what then happened to the blood. This has been given by historians the title of the 'lesser' circulation. It would have come to Harvey's attention in the course of his reading: it was a known opinion, but did not yet have the status of truth – partly because it gave an explanation to something which most anatomists did not see as a problem. For, in the conventional view, both the nutritive blood of the veins and the vivifying blood of the arteries were regularly consumed in the process of keeping the body alive: as all the blood was thus accounted for, there was no call to discover new pathways for it.

So the regular physician of the early 1600s took it for granted that the blood in the human body moves, but he believed that it moves in two different forms, in two kinds of blood-vessel, and for two different purposes. All of this, it must be noted, had been investigated and worked out in order to understand the functioning of the body of *man*, for this is what Galen the physician, who was its major author, had been concerned with, and he had been followed in this by all subsequent anatomists. Man has a heart with two ventricles or 'bellies' (and two auricles or 'ears'); man has a lung: this account deals with and ascribes a role to all these parts. As long as your anatomical investigations are limited to the body of man or have the primary aim of elucidating the workings of the body of man, then nothing is going to seem radically wrong with this Galenic account. But if, by contrast, you are going to investigate the detailed operations of the heart and its vessels in a creature *other than* man, then you may well see things differently.

This is just what Harvey did. For what he was investigating was not 'man' but 'the animal'. That is to say, the account he was trying to give of what the heart is for – what role it plays in the life of 'the animal', and why it is necessary for that life – had to be true of *all* hearts in *all* animals. It had to be true of hearts which have fewer chambers than the human one; it had to be true of the hearts of animals with cold blood as well as those with warm. Most important of all (as it turned out), it had to be true of the hearts of animals without a lung as well as those with. But, as Harvey found very early on in his research, 'whatever earlier people have said about systole and diastole – about the [alternating] motion of the heart and arteries – all these things they have related while eyeing the lung!' Harvey, however, was going to look at the heart and arteries *without* considering the lung. He has a novel object of study: it is the heart (the whole of it) plus the arteries. No-one since Aristotle had investigated this particular set of things.

Thus, at the very moment that he took up this set of organs to explore, Harvey was out on his own. His only predecessors in such inquiry were Aristotle and Fabricius, and eventually he was reduced to criticising even them. But because of the nature of his project, a great many things were visible to Harvey which were not readily visible to investigators interested only in man. For, as he was seeking an answer true of *all animals*, so he took as his experimental material as many kinds of animal as he could lay his hands on. It followed, for his project, that he would dissect the slow-beating hearts of fish, snakes and other cold-blooded creatures. Thus was he enabled to distinguish between the alternate movements of the heart and to come to believe that the tip of the heart beats against the chest-wall, and the blood is expelled from the ventricles, when the heart is hard and contracted (in our terms, when the heart-muscle contracts), not as a result of being inflated with blood. This reversed the customary view as to which state was diastole (expansion) and which was systole (contraction). Again, because he was interested in the *movement* of the heart in all animals, it followed that Harvey would constantly conduct vivisectional experiments, and so be able to see the movements of the heart more clearly as it was dying. And because he was interested in the 'heart and arteries' (not in the left-heart, arteries and lung), he was constantly inspecting and vivisecting animals without a lung, but with some other means of respiration. He was thus able to work out the relation of heart to lung, and to understand which parts of the heart were essential to it being a heart, and which merely served a particular mode of respiration.

Where a Galenic anatomist might have occasionally vivisected a

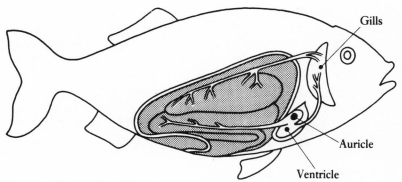

Figure 1. An instance of the heart without the lung: the fish.

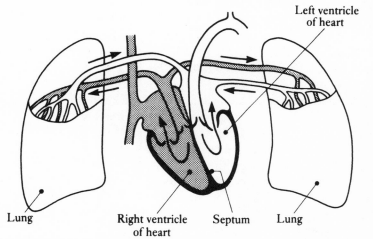

Figure 2. The 'lesser' circulation (right heart–lung–left heart) in the human, as suggested by Columbo.

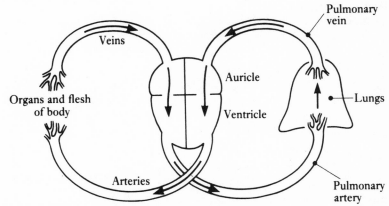

Figure 3. Schematised view of the full circulation in the human.

dog or pig to discover something or other about the functioning of the body of man, Harvey constantly vivisected all possible shapes, sizes and conditions of animal, because he was investigating the 'use and utilities of the motion of the heart in animals by the use of my own eyes (*per autopsia*)'. In other words, the very nature of his project brought to his attention things which no other anatomist was seeing. For a Galenic anatomist the heart of a snake tells him nothing about the heart of man; for an Aristotelian anatomist like Harvey, the heart of a snake is a particular instance of the very thing he is investigating: 'the heart' in 'the animal'. It was knowledge acquired from the investigation of the heart of 'the animal', which was the raw material of Harvey's whole sequence of inferences, and his whole argument.

It is not currently possible to reconstruct the precise sequence of experiments which led Harvey to discover the circulation of the blood, for the book he published in 1628 about the discovery was a carefully worked-out argument, not a laboratory notebook. He called his book *An Anatomical Exercise on the Motion of the Heart and Blood in Animals*, and wrote and published it in Latin. But we can listen here to the breathless way he describes the moment when it occurred to him that the blood circulated. Harvey has been arguing that the blood must pass from the right side of the heart to the left via the lungs (the so-called 'lesser' circulation):

Truly when I had often and seriously considered with myself and long turned over in my mind how great the quantity of blood was – evident partly from the dissection of living animals for experiment's sake, and from the opening up of blood-vessels and from many ways of investigating, partly from the symmetry and magnitude of the ventricles of the heart and of the vessels which go into it and out from it (since Nature, making nothing in vain would not have allotted that greatness proportionately to those vessels, to no purpose), partly from the beautiful and careful construction of the valves and fibres, and from the rest of the fabric of the heart, and likewise from many other things – viz., the abundance of blood passed through was so great, and the transmission was done in so short a time, that I realised that the fluid from the ingested food could not supply it: and indeed that we should have the veins empty, quite exhausted, and the arteries on the other hand burst with too much intrusion of blood, unless the blood did pass back again by some way out of the arteries into the veins, and return to the right ventricle of the heart.

I began to bethink myself whether it had a sort of motion as if in a circle, which afterwards I found to be true, and that the blood was thrust forth and driven out from the heart through the arteries into the flesh of the body and all the parts, by the beating of the left ventricle of the heart . . . and that it returns through the veins into the vena cava, and to the right ear of the heart . . .

What settled it all for Harvey, it seems, was the presence of little 'doors' (membranous flaps) in the veins: such things do not occur in the arteries, so what purpose could there be for them in the veins? Fabricius had actually discovered these 'little doors', but had thought their purpose was to slow the outward flow of the nutritive blood. Faced with his new dilemma about the route of blood, Harvey investigated them anew and discovered that they allowed only one-way flow – that they were *valves*. Their role had to be to prevent the blood, returning through the veins to the heart, from slipping back.

Thus did Harvey discover the circulation of the blood. But he had not been trying to discover whether the blood circulated, nor had he initially even been trying to confirm a hunch that it circulated. Harvey was engaged in a quite different enterprise, with its own goals and purposes. This was one reason why the idea that blood circulated dawned on him only slowly and reluctantly. But there was another reason too. Harvey was a willing follower of Aristotle in his whole approach. Aristotle the philosopher and investigator of 'the animal', provided Harvey with his method, his logical tools, his topic, his goals. Indeed Harvey conducted his anatomy in a philosophical spirit: like a good philosopher he was looking for *causes*, trying to give accounts which explained why 'the things themselves' are as they are. The problem that arose from Harvey's devotion to Aristotle was that Harvey could not be totally happy with his new discovery, because he could not find out the most important thing about it: he could not discover *why* – for what purpose – the blood circulated. For a philosophical anatomist, as Harvey was, this was the greatest disappointment: he had to settle for 'speculating' and 'likely reasons'.

Harvey's discovery was not greeted with immediate belief. Many people thought it was absurd, and others took it as a threat to their understandings of how the body worked. Controversy went on for years. John Aubrey, the gossipy antiquarian, heard about it from Harvey himself:

I have heard him say, that after his book of the *Circulation of the*

Blood came out, that he fell mightily in his practice, and that 'twas believed by the vulgar that he was crack-brained; and all the physicians were against his opinion, and envied [had ill-will towards] him; many wrote against him. With much ado at last, in about 20 or 30 years time, it was received in all the universities in the world; and . . . he is the only man perhaps that ever lived to see his own doctrine established in his life-time.

Harvey's discovery of the circulation of the blood was an accidental by-product of his attempt to revive and reinstate in seventeenth-century England the anatomical enterprise that Aristotle had conducted in Athens twenty centuries before. This greatest of modern discoveries about the body was made by a man trying to do anatomy like an ancient Greek philosopher. Only an Aristotelian anatomist could have thrown up the particular set of phenomena which required the hypothesis of the circulation of the blood to explain them: and only an Aristotelian anatomist did. Even though in Harvey's day there were non-Aristotelian and even anti-Aristotelian philosophers, none of them was asking the kind of questions, and of the right object in nature ('the animal'), which could have led them to think that the blood might be circulating. Such questions made sense only to Aristotelians. The discovery of the circulation of the blood was exclusively (and necessarily) the product of an Aristotelian way of thinking and investigating.

All this may seem a bit strange. For we would instinctively expect this great discovery to have been a product of a deliberate attempt to release men's minds from their bondage to the teachings of the 'ancients', and in particular from the teachings of Aristotle. Given that this discovery was made in seventeenth-century England, we would also expect it to be associated with the 'mechanistic' way of thinking then associated (in different forms) with the work of René Descartes, Robert Boyle and Isaac Newton. 'Mechanistic' explanations of natural phenomena are ones which seek to explain events primarily in terms of the motion of the smallest particles of inert matter: it is the approach of which we are the heirs, and which underlies our modern science. We are so imbued with this way of thinking, that the temptation is for us to reconstruct this discovery so that it accords with a mechanistic view, by (for instance) stressing the one instance of measurement that Harvey uses in his argument, and the analogy he makes between the action of the heart and that of a pump. And, because it seems so obvious to us that the blood circulates – it seems to us that it was a fact just waiting to be discovered – we would expect Harvey to have been the final person in a tradition of people *looking for* evidence and proof of the circulation. But all this would be wrong. For Harvey, the body did not act like a machine in any way: it was alive, and all its activities and functions were those of life. For him the beat of the heart was a vital, not a mechanical, phenomenon, and the blood itself was alive.

In his own day Harvey's approach, though only newly brought back into practice, was very old-fashioned. He looked to Aristotle as his primary authority, and had no time at all for such people of his own day as Descartes, who were promoting the 'mechanical philosophy'. Indeed Harvey's anatomical practice was as old-fashioned as his politics. Throughout his life he was a loyal supporter of the monarchy, never wavering even in the darkest days of the Civil War and Commonwealth. He dedicated his book announcing the discovery to the King, his patron and employer, Charles I. It was not just a matter of flattery that Harvey's opening words there describe the king in his kingdom as being like the heart in the body: 'Most serene King', he wrote, 'the heart of animals is the foundation of life, the ruler of each of them, of the microcosm it is the sun from which all *vegetatio* derives, all vigour and strength emanates. The king is equally the foundation of his kingdoms, and of his own microcosm he is the sun, of the State he is the heart from which all power emanates, every favour originates.' It is clear that when, as a loyal Aristotelian, Harvey first chose to study the heart in the animal, he was also, as a loyal monarchist, choosing to study it as the king of the animal body.

ISAAC NEWTON:
The Mathematical Key to Nature
A. Rupert Hall

Isaac Newton (1642–1727) was a man of many facets, each of which reflected his times. There was the Newton who was the greatest mathematical physicist of the seventeenth century – the man who brought order to the 'New Science' which advanced in so many unforeseen directions. But there was also Newton the theologian and Newton the alchemist – reminders that it is naïve for us to assume that science in the past was rigidly divided from other forms of belief. And there was Newton who was knighted and made Master of the Mint – sure signs of early links between science and state. And, not least, there was Newton the President of the Royal Society. Chartered by King Charles II, the growing importance of the Royal Society is an indication of how science was starting to organise itself at that time and recognise that its own advancement depended upon a strong sense of community.

The long life of Isaac Newton embraced the execution of Charles I and the restoration of Charles II, the Glorious Revolution and the Whig Ascendancy. He actively resisted James II, sat in the Convention Parliament, and faithfully served both Queen Anne and George I. Newton never saw London before the Great Fire (1666) but was concerned in the reconstruction of the city by Wren and his successors. As an authority on astral navigation consulted by the Admiralty, Newton may even be said to have had a tiny share in Britain's grand maritime expansion.

Newton moved to London in 1696 as, first, Warden, then Master of the Royal Mint where he laid a sound monetary basis for Britain's eighteenth century prosperity. He was knighted in 1705 having begun a long, autocratic presidency of the Royal Society less than two years before. The first and prodigiously creative half of his life Newton spent in Cambridge, where he was appointed Lucasian Professor of Mathematics at the age of twenty-seven. Before he was thirty he was internationally celebrated for his experiments on light, but it was above all as a mathematician that Newton earned his unparalleled place in history. His exceptional mental powers first manifested themselves in his early mathematical notes; his swift ability to solve mathematical problems enabled him to lay the basis of classical physics and, in so doing, create the modern world-view. Newton retained these powers to an old age, disposing during a single evening in 1697 of a challenge to all the mathematicians of Europe, which only two others mastered. His rivalry with Leibniz over the invention of the calculus brought vigour as well as acrimony to his later years: longevity assured Newton the victory.

But Newton was not only outstanding as a mathematician. He was versatile, precise, imaginative and assiduous in experiment. His regular chemical researches, in a laboratory by the Great Gate of Trinity College, occupied him for twenty-five years. He made the first practical reflecting telescope with his own hands and devised a form of reflecting sextant. He was, besides, a natural philosopher, theologian, classical scholar and the most thorough student of alchemical authors of his time; had Newton not been the greatest mathematician of the age he would have been its greatest scholar. Newton attributed his success in these diverse pursuits to long hours of work, intensely concentrated attention, and relentless curiosity; he did not emphasise the remarkable intellectual capacity that penetrated and directed his long hours of study and writing – a capacity as evident in his conduct of government business as in science.

'There could be only one Newton', others said later, meaning that he had been fortunate in the moment of his birth, that is when the pieces of the puzzles of the universe had been assembled, but recognising also that, to see the picture, surpassing originality was required. No one else came close to Newton in solving this puzzle. In Italy the mathematical science of motion had been created by Galileo and his successors; in central Europe astronomers (Copernicus, Tycho Brahe, Kepler) had elucidated the pattern of motions in the heavens; everywhere new ideas of matter, its properties and its interactions were taking shape. As yet, however, in each of these fields of activity there was confusion and obscurity. None was complete and none was related logically to any of the others. In the past it might have been enough to affirm that it was 'natural' for planets to revolve in circles – now, the elliptical orbit demanded explanation of the planets' approach towards, and recession from, the Sun. Moreover, as each planet revolved in its elliptical orbit its speed varied in a particular way (discovered by Kepler) while the *average* speeds of all the planets were again related one to another in a manner also discovered by Kepler. All these various relations demanded explanation: in brief, as soon as the long-established system of rigid planetary spheres was done away with, some physical mechanism to account for the newly discovered laws of planetary motion had to be devised. Johannes Kepler supposed that a mutual magnetic attraction might bind the celestial system together. A generation later, René Descartes imagined the Sun at the centre of a whirlpool of aetherial matter that swept the planets around. Both these explanations were vague, non-numerical and unverifiable. They provided at best philosophies of nature (far-reaching in Descartes' case) unrelated to the actual measurements and observations of astronomers. Since the physical analysis of the simplest motions of ordinary bodies was still very primitive it is not surprising that the planetary motions defied physical analysis altogether.

The basis for such an analysis was constructed by Galileo (*Discourses on Two New Sciences*, 1638) with clear definitions of uniform and accelerated motion and the first solution of motion in a curve (the parabola of projectiles). About mid-century the properties of the cycloid were established (this is the curve traced by a point on a rolling circle), and soon after Christiaan Huygens extended the analysis of movement to the swinging pendulum. These were the problems with which the youthful Newton began, as a Cambridge undergraduate. In old age he penned a famous (and, broadly speaking, accurate) narrative of the course of his earliest mathematical and scientific inquiries:

In the beginning of the year 1665 I found the method of approximating series . . . and in November had the direct method of fluxions (calculus) and the next year in January had the theory of colours and in May following I had entrance into the inverse method of fluxions (i.e. integration, in the calculus). And the same year (1666) I began to think of gravity extending to the orb(it) of the Moon and . . . I deduced that the forces which keep the planets in their orb(it)s must (be) reciprocally as the squares of their distances from the centres about which they revolve, and thereby compared the force requisite to keep the Moon in her orb(it) with the force of gravity at the surface of the Earth and found them answer pretty nearly. All this was in the two plague years of 1665–1666. For in those days I was in the prime of my age for invention and minded Mathematics and Philosophy more than at any time since.[1]

[1] The calculation about the Moon had been scribbled on the back of an old lease granted by his mother.

Inevitably, after half a century Newton's recollection of what he had accomplished at home in Lincolnshire, when Cambridge University was closed by the plague, smacks of a firmer and more positive grasp than he actually possessed at the time. Nevertheless, he had his methods of fluxions and of infinite series fully worked out on paper only five years later (1671). Before this, Newton had 'written up' his optical investigations, the subject of his first professorial lectures at Cambridge (1670). The preliminary account of them that appeared in the Royal Society's *Philosophical Transactions* in 1672, soon after Newton had been elected a Fellow, occasioned great excitement and some scepticism.

His work on mechanics and the theory of universal gravitation took much longer to mature, not least because the explorations made in 1665–66 seem to have been set aside for many years while Newton pursued first his mathematics, then his optics (debate about his revolutionary ideas went on into the mid-1670s) and thirdly his chemical experiments. He had studied (and rejected) Descartes' formulation of the phenomena of moving bodies, and probably clarified his own ideas early. He obtained a clear and correct principle for interpreting the rebounding of colliding bodies and independently solved the problems of circular motion and the pendulum (first resolved in print by Christiaan Huygens in 1673). Knowing how to calculate the centrifugal acceleration of a circularly revolving body, knowing, too, Kepler's Third Law of planetary motion that (T^2/r^3) is a constant ratio within the solar system,[2] Newton found that the acceleration of the Moon in its orbit away from the Earth was 'pretty nearly' equal to the opposite acceleration of a heavy body towards the Earth at the Moon's distance. This conclusion relied on the assumption that terrestrial gravity decreased in the ratio of the squares of the distances. Somewhat later Newton satisfied himself that the centrifugal accelerations of all the planets in their orbits about the Sun – taken as circular – followed the same inverse-square pattern. But Newton was as yet scarcely aware that he had in his hands the key to a grand system of celestial mechanics, and was far from attaining to mastery of its full conceptual and mathematical structure. For at least ten years more he continued to believe that 'together with the force of gravity there might be a mixture of that force which the Moon would have if it was carried along in a Cartesian vortex' (to employ the confused expression of Newton's heir, John Conduitt). There was also the problem of proving that the inverse-square ratio

[2] T is the periodic time of any planet's revolution, r its mean distance from the Sun.

could as well apply to elliptical orbits as it could to circular ones.

There was, therefore, an element of biographical paradox in the fact that Newton's first and greatest book, on which his fame was chiefly founded, took as its subject this science of mechanics to which he had devoted comparatively little of his time since 1666. Newton owed the revival of his personal interest in the mechanical analysis of planetary motion, and hence the discovery that its development was within his power, to three contemporaries: Robert Hooke, John Flamsteed and Edmond Halley. (Only the last-named was to remain Newton's friend and partisan: sadly, both Hooke and Flamsteed became bitter enemies, though the latter had provided many of the astronomical data used by Newton in constructing and verifying his system.) An exchange of letters in 1679–80 with that experimentalist of genius, Robert Hooke forced Newton to reconsider circular motion and to devise a geometrical proof that revolution in an ellipse is a necessary consequence of an inverse-square centripetal force retaining the revolving body in an orbit, the force being directed towards one focus of the ellipse.[3] This truth was first proposed in their discussion by Hooke, who asked Newton to prove it, and Newton did so. This success, of which Newton left Hooke ignorant, was a great encouragement (it is immortalised in the first proposition of the *Mathematical Principles of Natural Philosophy*, 1687) yet many equal or greater mathematical difficulties were to be encountered in the progress of the book. A year later, a correspondence with the Astronomer Royal, John Flamsteed, about the motion of comets convinced Newton that these capricious bodies swing in a tight arc around the Sun, indicating the action of a powerful solar centripetal force. Thirdly, Edmond Halley, who was years afterwards to be Flamsteed's successor at Greenwich, rode expressly to Cambridge in August 1684 to put the precise question to Newton: what shape would the orbit of a planet have, if the planet were constantly drawn towards the Sun by an inverse-square force? Newton was able to reply at once: 'An ellipse: I have proved it.'

Thereafter, writing first for Halley's satisfaction, then for that of the Royal Society and so to enlighten the whole world, the incomparable *Mathematical Principles* were begun, to be published less than three years later in June, 1687. Though the fall of an apple in 1666, while Newton sat 'in a contemplative mood' in the

---

[3] The circle is a special, zero-eccentricity, case of the ellipse; Newton had not yet discovered that the parabola is another special case, the ellipse of infinite eccentricity.

orchard of his birthplace at Woolsthorpe (Lincolnshire) may have suggested to him the existence of 'a drawing power in matter', the transformation of this vague idea into a mathematically articulated scientific theory buttressed by an enormous variety of precisely directed confirmatory evidence was a lengthy process, at first slow and almost unconscious, then at the last (after August 1684) an amazing work of swift, continued and concentrated effort. The essence of the future *Mathematical Principles* was already contained in a sketch completed before the end of 1684.

The *Mathematical Principles* with their clear and convincing foundations, their subtle and elaborate mathematical development and their extensive application to the most majestic phenomena of nature (the celestial bodies, the shape of the Earth, the tides, the motions of bodies in fluids and of sound in air, the laws of hydraulics and pneumatics) outshone by any comparison everything in man's renewed exploration of Nature since the Renaissance. The book created a new model for future science just as Archimedes had done two thousand years before, by proving that the behaviour of floating bodies was governed by mathematical principles. No reader trying to follow Newton's mind has found the task easy, though the geometrical structure of the *Principles* was neither novel nor recondite. In the following Age of Enlightenment the mass of readers knew Newton – who influenced that age so profoundly – indirectly through popularisations in many languages, and directly through the more approachable *Opticks* (1704). Popularisers, including Voltaire, brought out strongly the contrast between the aether-filled universe of Descartes, wherein everything was explained by the impact of different kinds of invisible particles upon solid and visible bodies, and the empty universe of Newton, wherein the phenomena we experience are caused by the action of forces (cohesive and attractive, or dispersive and repulsive) acting between the particles – or atoms – of which bodies are made, and their compounded molecules.[4] It was by this idea of *physical force* that Newton utterly remade the mechanics and the general physical thinking of the age. Many distinguished contemporaries found the idea of forces, imperceptibly active within the emptiness of space, hard to grasp; mathematically, it was to be refined by the French mathematicians of the eighteenth century; conceptually, it was to be enriched and subsumed within a greater intellectual framework by the physicists of the nineteenth.

[4] Newton would have said corpuscles.

Although the gravitational force is the major topic of Newton's masterpiece it is certain that he envisaged the existence of other types of force, including a chemical force (or forces) and an optical force operating between ordinary matter and the matter of light. Rejecting the then usual explanations that light is either a pressure or a vibration (wave) transmitted from source to eye through the omnipresent aether, he believed light to consist of the almost infinitely swift passage of almost infinitely minute particles through (possibly empty) space. But because Newton was forever unable to provide a satisfactory demonstration of such a theory of light, *Opticks* is both experimental and phenomenalistic, its chief contentions being independent of any particular theory of the nature of light. Newton's fundamental experimental discovery, made in 1664 or 1665, was that white light is not simple and homogeneous but rather composed of coloured constituents visible as separate rays in the spectrum. Each constituent ray he found to be indivisible and characterised by a particular mathematical property (corresponding to wave-length in our theory). Refraction, making the colours visible in the spectrum, is, as it were, an analysis of white light, which can be recompounded, while pigment colours are produced by the selective absorption, in the surface upon which white light falls, of one or more of the primitive colours. (This is another form of interaction between matter and light.) Newton found later that he had to add further complications to his basic idea in order to deal – inadequately, it must be said – with such later discovered effects of light and colour as diffraction and double-refraction.

His idea of light – both his 'doctrine' of the primitive coloured constituents to which he was firmly committed, regarding this as an inescapable consequence of his own experiments, and his own strongly preferred (but never completed, nor demonstrated) material or particulate hypothesis – was in strong contrast to others widely held.

The oldest explanation of colour is that it represents a diminution of light, black being an absolute privation. Somewhat analogously, many of Newton's contemporaries (Huygens and Hooke among them) thought that refraction or reflection, causing colour to be seen, was a process of physical *modification* of white light, the alteration being reversible. Newton's prime discovery was that a pure red ray (for example) cannot be made to yield white or any other colour, but it took more than fifty years to gain general acceptance. The material theory of light was never, in fact, universally accepted, and was the first element in Newton's scientific legacy to be rejected. From about 1830 a re-formulated wave-theory was accepted everywhere.

Historians have often spoken of the 'Newtonian synthesis', meaning his successful and fertile amalgamation of astronomy with the science of motion. If with less enduring effect, Newton also brought about a partial reconciliation of physics with the science of motion, and of all these branches of science with fundamental conceptions of the nature of properties of matter. Thus, as the title of Newton's greatest work indicates, he effected a synthesis of mathematics and natural philosophy, a synthesis of which the Greeks had despaired and towards which Galileo and Kepler had only pointed the way. Before Newton, the mathematician and philosopher had undertaken separate tasks (this is very evident from comparison between the various writings of Descartes, who never succeeded in uniting them), that of the philosopher being regarded as both broader and more fundamental. After Newton, it was clear that mathematics and experiment conjointly were the basic keys to the formulation of a natural philosophy. The partial disclosure of a grand intellectual vision that readers found in the series of *Queries* that Newton appended to the (unfinished) text of *Opticks* proved of lasting fascination. The natural-philosophical message they contained stemming from the conviction that

> God in the Beginning form'd Matter in solid, massy, hard, impenetrable, movable Particles ... that these Particles have not only a *Vis inertiae*,[5] accompanied with such passive Laws of Motion as naturally result from that Force, but also that they are moved by certain active Principles, such as is that of Gravity, and that which causes Fermentation, and the Cohesion of Bodies. (Query 31)

was actively developed in the Anglo-Saxon world – for example, it led Stephen Hales to his experiments on plant physiology and Benjamin Franklin to his investigation of electricity. On the other hand Newtonian mechanics was brought to its supreme development by continental mathematicians. If the brilliant French writer and *philosophe* Voltaire was outstanding among popularisers of Newton, it was Maupertuis who began the naturalization of the Newtonian system in France, and who promoted the scientific expeditions which, by geodetic measurements, verified Newton's theoretical calculations of the Earth's oblate form. Later French

[5] i.e. 'force of inertia' – the momentum of a moving body.

mathematicians, notably Legrange and Laplace, employing the differential and integral calculus of Newton's great rival Leibniz, perfected the Newtonian system in many respects, as by the proof that the universe is essentially stable and not (as Newton feared) subject to an increase of disorder and loss of motion that only God could remedy. They also effected a reconciliation of certain Leibnizian principles in mechanics with those already established by Newton. Philosophy, declared Newton in a moment of exasperation, is a litigious lady. Such was his own experience: optics, astronomy, mechanics – all his published investigations involved him in troublesome controversies. But not publishing was as productive of acrimony as publishing. In the early 1670s Newton contemplated but failed to complete a printed account of his mathematical studies. His hesitancy allowed Leibniz to anticipate him with respect to the 'direct method of fluxions' (1665) by printing Leibniz's own differential calculus (1684).[6] After inserting a brief note into the *Mathematical Principles* three years later, Newton at last published two mathematical essays with *Opticks* in 1704. By this time his priority wrangle with Leibniz concerning the discovery of fluxions/calculus was in full flow and (of greater long-term significance) the development of the Leibnizian calculus in the hands of such able mathematicians as the brothers Bernoulli was proceeding swiftly.[7] The essays, and mathematical works of Newton, printed subsequently, were to have negligible creative effect on the development of the subject.

Thus the great richness and originality of Newton's mathematical discoveries was for forty years denied to all but close friends. Similarly, outside mathematics and physical science Newton was largely successful in concealing his beliefs and wide interests. Robert Boyle (1627–91) and the philosopher John Locke (1632–1704) knew of his interest in alchemy, which they shared, but a brief paper *On the Nature of Acids* was the only indication and fruit of Newton's chemical preoccupations to appear in his lifetime other than the examples of the operation of a chemical binding-force that he inserted into the *Queries* in *Opticks*. The reams of transcripts from alchemical authors made by Newton,

---

[6] Already outlined for Newton's benefit in a private letter of 1676.
[7] Roughly speaking, fluxions and calculus are equivalent but distinct algorithms (or technical systems) for handling quantities that are continuously varying in some manner, as the speed and distance with respect to time in a non-uniform motion. The opening up of these new mathematical horizons provided vast new powers in studying the properties of curves and so of innumerable problems in mechanics and physics.

like his own attempts to reduce their esoteric language and fanci-fully symbolic descriptions to a coherent order and subject them to experimental exploration were otherwise to remain unknown down to recent times. Newton was not searching for man-made gold but for insight into the structure of matter; after thirty years of trials he assured Locke that there was an insuperable argument against the possibility of chemical transmutation. Another long concern of Newton's was the early history of man; like alchemy it offered the possibility of 'decoding' hidden meanings, which inevitably involved the calculation of a date for the creation of the world. An abridgement in French of Newton's *Chronology of Ancient Kingdoms Amended* were clandestinely printed two years before his death, much to his annoyance. The book illustrates Newton's tireless pursuit of detail, and a 'scientific' attempt – less than wholly successful – to obtain absolute dates for historical events from astronomical data. The *Observations upon the Prophecies of Daniel and the Apocalypse of St John* were not printed until six years after Newton's death. The theological writings that have aroused greater interest in recent decades are those in which Newton questioned the rectitude of the Roman Trinitarian tradition and demonstrated his personal predilection for a unitarian interpretation of Christianity. These remained in inviolate secrecy until the nine-teenth century. Newton was well aware of the catastrophe awaiting those, like his chosen mathematical successor at Cambridge, William Whiston, who were imprudent enough to make public any scepticism over or about the Three Persons of God.

Caution, prudence and reserve were natural elements in New-ton's character. Any natural tendencies to exuberance in him were, like the slightest leaning towards convivial dissipation, soon outgrown. Springing from the lowest stratum of the landed gen-try, his father unable to sign his name, Newton had met with little family understanding of his own intellectual interests: it is always less easy to live with genius than to admire it posthumously. Advancing far from his roots in a farm (while preserving his manorial privileges) Newton nevertheless retained some peasant attributes: though he became a man of wealth and position he remained careful with his money and inwardly insecure; his vul-nerability was concealed by an appearance of authority, even arrogance. Such men as Hooke, Flamsteed and Leibniz by whom Newton believed himself to have been wronged were never for-given, and Newton was unscrupulous in obtaining what he took to be due redress. On rare occasions the tensions of an unhappy childhood, of social inferiority and intellectual frustration burst

86

out in terrifying rages. To his friends, to foreign visitors, to his government masters Newton was a warm host, a good Lincolnshire man, the affable philosopher. His most appealing personal trait, deeply rooted in his sense of family, was his constant affection for his half-niece, the beautiful Catherine Barton, friend of Halifax and Swift.

During his lifetime, even enemies recognised Newton's surpassing abilities. Interest in his personality was keen during his last, public years, as the man with all-too-human flaws became the hero of almost divine powers:

> *God said: Let Newton be!*
> *And all was Light.*

London became the scientific capital of Europe, but 'Newtonian' rather than 'English' science was to furnish the scientific basis for the Enlightenment. Newton's role as one legislator of a new, rational order of society would perhaps have been as little welcome to him as Laplace's exclusion of God from the Newtonian universe. As happened many times, the conservative by temperament initiated a revolution in thought whose limits are not yet to be guessed.

JOSEPH PRIESTLEY:
Science, Religion and Politics in the Age of Revolution
John Christie

Joseph Priestley (1733–1804) is an extraordinary example of a man
of phenomenally broad talents. In the history of science, he is best
known as a chemist, indeed as perhaps the man who 'discovered'
oxygen (a contentious accolade since Priestley always denied that
oxygen existed, and hence can hardly be said to have discovered it).
But he saw himself as a general 'natural philosopher', who also
made great contributions to fields such as the study of electricity.
Priestley wrote history books and spent time as a teacher. He was
also a theologian and a protestant dissenter, who became
embroiled in deep controversy with the Anglican establishment.
Moreover, he was a political activist – a radical – who was finally
forced to flee to America after being persecuted at home for his
extreme views.

Joseph Priestley (1733–1804) is probably the best-remembered man
of science to work in Britain in the latter half of the eighteenth
century. There were others, such as Henry Cavendish and Joseph
Black, whose scientific attainments might be rated more highly, but
none have remained in common memory as persistently as Joseph
Priestley. Within history of science, he is usually, if somewhat
misleadingly recalled for his discovery of oxygen, and for his
subsequent opposition to the new chemistry associated with
Antoine-Laurent Lavoisier. But he also maintained a high public
profile as a leading religious Dissenter, and as a champion of
political liberty in the age of the American and French Revolutions,
gaining a notoriety which eventually resulted in the destruction of
his house and laboratory in Birmingham in July 1791 by a patriotic
'Church and King' mob. He cannot be ignored by historians of
science, religion or politics. Other branches of history also require
to take account of Priestley. He was, for example, a great educator
and a considerable philosopher. No activity he engaged in was left
unaltered by him. What Priestley touched, he changed. The
historians of science who have written this book may even count

Priestley as one of their own intellectual ancestors, for his books on the history of electricity and optics were pioneer works in the history of science.

All this makes Priestley a difficult figure for our own time to comprehend fully. His range of interests and abilities was poly-mathic, unusual even for the eighteenth century. Few people nowadays in our age of specialisation have the range of skills and sympathies necessary for a complete and thorough understanding of this complex and controversial man. To this sort of difficulty we could add another, namely that of the contradictory nature of Priestley's career. In science, although he was a notable discoverer, Priestley devoted much of his effort to the conservative enterprise of opposition to certain crucial features of Lavoisier's chemistry. In politics, however, he was for his time a committed radical, deeply opposed to basic aspects of the political and social systems within which he lived and determined to change them. At a more general level, Priestley devoted himself to certain universal ideals, such as liberty, progress and peace, but his very pursuit of those ideals created a life of conflict and confrontation. From a certain perspective, Priestley's career is one long sequence of disagreements and debates, a sequence culminating in his leaving England for what he hoped would be the freer air of America.

That was the final major move of a highly mobile life. Priestley was born on 13 March 1733 at Birstal in the environs of Leeds. His mother died when he was six, and his father, a cloth-dresser by trade, later arranged for Joseph to be brought up by his aunt in Heckmondwike. He attended local schools, making good educational progress despite a tubercular illness. The dominant feature of his family life, local culture and education was religion: his upbringing was thoroughly pervaded by the Calvinist variety of Nonconformism, and the direction of his education toward Latin, Greek and Hebrew indicated a preparation for the ministry. This preparation continued at the new dissenting academy at Daventry. Priestley's time there turned the hard-working student into an emergent scholar with a formidable array of talents, great energy and an omniverous spirit of inquiry. To his language he added biblical criticism and theology, and also engaged himself with contemporary science and philosophy.

After Daventry, Priestley's career took him through contrasting environments as teacher at the Warrington Academy, minister in Leeds, tutor and librarian to the Earl of Shelburne's family, and private citizen and minister in Birmingham. More significant than the patronage of the great, such as Shelburne, was the constant material, moral and intellectual support given Priestley by like-minded men. These networks of friends, composed of men and women of similar religious and political commitments, including industrialists and fellow men of science, gave Priestley much assistance throughout his adult life. When, for example, he moved to Birmingham in 1780 after his time with Shelburne, it was through the considerable financial aid of such people.

The picture we have of Priestley then is one of a man whose restless energy and variety of commitments created a less than settled life, and whose often outspoken character generated hostility from those he opposed and those who felt threatened by Priestley's values; nonetheless he was equally capable of arousing strong friendship and loyalty among those who knew him well. To this picture we need to add certain features important for an understanding of Priestley's specifically scientific career. Relatively early on at Daventry, he gained the invaluable habit of writing much and writing quickly, a habit which enabled him to maintain a high rate of literary production for long periods of time. From the point of view of his intellectual development, religion must once again be considered a key factor. Although he did not remain a strict Calvinist, he certainly remained a devout and active believer of a certain kind. The eighteenth century certainly witnessed attempts to secularise the whole field of intellectual enquiry, including the natural sciences, but Priestley was no secularist philosopher of the Enlightenment, and indeed wrote strongly against certain trends of irreligion in philosophy. By contrast, he remained committed to a distinctive brand of religious belief. He was a highly rationalist Unitarian, denying the Three Persons of the Trinity, for whom religion and the practice of science were not opposed, but co-operative activities. He could even maintain the position that, properly understood, Holy Scripture and the findings of rational enquiry were perfectly compatible. As Priestley perceived it, the world of nature, in its general frame and in its details, was divinely ordered and created. Men were placed within this divinely ordered creation not just to use it for their own purposes, but to perceive, understand and act upon the intentions of a benevolent God toward humanity. Set within this pattern of belief, natural science came to have two essential and related functions for Priestley. Firstly, proper understanding of nature, gained through scientific enquiry, led men to reinforced belief in God, for God's benevolence was displayed in the processes of nature. Secondly, through the increase of knowledge brought about by natural science, the historical development of humanity was substantially furthered. Unlike many of his British contemporaries, Priestley was a perfectibilist: that is, he believed in the possibility of the continual moral and material betterment of humanity through science, education, political reform and true religion, towards an ultimate state of perfection, this state being the destined goal of humanity by the intention of God, who had created man for this end.

As far as Priestley was concerned, therefore, science was by no means an isolated activity, unrelated to his other interests, some kind of relaxation from more demanding commitments. It was instead a vital part of a unified and, indeed, visionary programme for the whole human race, and it is important for us to realise these levels of motivation operating to further his scientific researches. He could even give scientific research a direct and provocative radical political edge. Priestley wrote the following passage, in his *Experiments and Observations on Different Kinds of Air*. From it we can gain a sense of the way in which science, religion, politics and progress were all part of one endeavour for him:

> The rapid process of knowledge, which, like the progress of a wave of the sea, of sound, or of light from the sun, extends itself not in this or that way only, but *in all directions*, will, I doubt not, be the means, under God, of extirpating *all* error and prejudice, and of putting an end to all undue and usurped authority in the business of religion, as well as of *science*; and all the efforts of the interested friends of corrupt establishments of all kinds, will be ineffectual for their support in this enlightened age; though, by retarding their downfall, they may make the final ruin of them more complete and glorious . . . And the English hierarchy, if there be anything unsound in its constitution has equal reason to tremble before an air pump, or an electrical machine.

No doubt this gives us Priestley in fine polemical vein, but it would be wrong to read such passages from his work as simple and deliberate hyperbole. Priestley shared with many contemporaries, both British and European, the idea that knowledge itself was not just a progressive force in the world, but was a liberating force for the world, able to free people from the shackles of ignorance, superstition, prejudice and suppression. Our own latter-day sophistication and pessimistic cynicism concerning the pro-duction of scientific knowledge and the uses to which it can be turned finds few strong echoes in the eighteenth century, with the exceptions perhaps of Jean-Jacques Rousseau and Jonathan Swift. We need not see Priestley as hyping up science when he envisaged the aristocrat trembling as he confronted an air pump. Edmund Burke, England's principal conservative polemicist in the period of the French Revolution, used imagery drawn from Priestley's own chemistry of gases to promote fears of revolution in the 'English hierarchy'. Similarly, Priestley's monistic, materialistic view of nature of the human mind could provoke alarmed reaction from the natural philosopher John Robison, who saw Priestley's work as part of a European philosophical conspiracy to overthrow good government, religion and morality. Science had its political meanings and uses for both radicals and conservatives at this time, so that the idea of an air-pump striking terror in the hearts of aristocrats was by no means as ludicrous in the latter decades of the eighteenth century as it might now appear.

Mention of air-pumps and electrical machines was itself by no means incidental for Priestley. He regarded the ability to construct such machines – both instruments of the first importance for scientific experimentation in the eighteenth century – as vital for anyone who seriously meant to pursue science. He himself gained the skills of an instrument-maker and constructed some of the apparatus with which he worked. He emphasised the importance of instruments for scientific teaching and used them consistently in his own educational practice. When his private Birmingham laboratory was destroyed in 1791, its equipment was valued at more than £600, a figure which compares very favourably with valuations of public laboratories of the period. This side of Priestley's science deserves especial emphasis. A casual acquaintance with his better-known works can result in the impression of a somewhat haphazard experimenter whose real intellectual *forte* was highly abstract philosophical speculation rather than practical science. He was undoubtedly a talented philosophical thinker, as his arguments with the Scottish philosophers David Hume and Thomas Reid demonstrate, and his views on the nature of physical matter were the product of much metaphysical speculation. But we nonetheless miss an essential aspect of his science if we rest with the view of Priestley outlined above, of half-planned experimentation surrounded by metaphysical fog. The care he devoted to instrumentation, both for teaching and in building up his own collection, shows a strongly practical conception of scientific investigation. He thought of the English political and religious hierarchy as threatened not by metaphysics, but by machines, and history did not prove him wrong.

Priestley's first great scientific interest was in the study of electricity. By the middle decades of the eighteenth century, electrical science, along with the science of heat and chemistry, were notable growth points within the field of physical science. It was in the 1750s that the chemist Joseph Black furthered the chemical understanding of 'fixed air' (carbon dioxide), and then produced the discovery of latent and specific heat. The preceding four decades had seen much theorising and experimenting in electrical science by men such as Francis Hauksbee in London, and Abbé Nollet in France. By the mid-1750s the Pennsylvanian Benjamin Franklin had completed his highly influential electrical experimentation and theory of the nature and action of electricity. Priestley undertook his own experimentation, but it was not a concerted attempt to further the science systematically. He experimented mainly to familiarise himself with electrical science, a considerable task which took him an astonishingly short time given the practical experimentation he was undertaking. The book which resulted, *The History and Present State of Electricity, with Original Experiments*, published in 1767, took him little more than a year to complete, was notable for its emphasis on the progressive nature of experimental knowledge, and included his own experimental discovery of the high conductivity of charcoal.

This kind of electrical experimentation was not simply electrical, but was connected to a fundamental scientific interest of Priestley's, one so long-standing as to be traceable back to his childhood, when, at the age of eleven, he had imprisoned spiders in bottles to ascertain how long they would survive without a replenished supply of air. The charcoal discovery was actually part of an experimental attempt to restore air which had passed over burning charcoal or through lungs. Such air, as was well-known, no longer supported the flame of a burning candle, and Priestley was seeking to restore this property by electrifying the air.

Both the early spider experiment and this electrical experiment of May 1766 indicate his abiding interest in the category of good air: air, that is, which will support basic processes such as breathing and burning. Much of his work on the chemistry of gases was not simply to do with the discovery of new gases and their distinguishing properties. It was to do also with 'vitiation', how air became 'spoilt' in various chemical processes, how its goodness might be restored, recognised and even measured. He produced, during the 1770s, techniques and instruments for just this purpose.

It is worth reflecting briefly on the implications of Priestley's basic stance towards air. 'Goodness', from one point of view, could be simply a functional property: does a given sample of air support respiration and burning? But 'goodness' for someone such as Priestley, committed to a view of nature as a harmonious, beneficial system created to support life by a benevolent Creator, was also a moral quality. For us, the idea that nature is a system not just of physical properties but also of moral qualities is an alien concept; but for Priestley, convinced of the Creator's benevolent intentions and eager to demonstrate them, such an idea would have been a reflex assumption, built into the way in which he perceived and studied natural processes. This should be borne in mind as we follow Priestley in his investigations of the chemistry of air.

The scene of these investigations was Leeds, where Priestley had moved in September of 1767. By this time he had been elected a Fellow of the Royal Society of London and some of his best known works had appeared or would shortly appear, including the *Essay on Education*, the *Institutes of Natural and Revealed Religion*, and the *Essay on the First Principles of Government*. His time in Leeds, however, is chiefly memorable as the period in which he prosecuted his experimental enquiry into different kinds of gases.

Why did Priestley undertake this task? We have seen one reason already, his pre-established interest in the goodness of air. To this can be added two further factors. Firstly, there was an opportunity to be seized. In Leeds, Priestley lived close by a brewery, whose fermentation process offered an abundance of available 'mephitic' or fixed air, and he accordingly took the chance to perform experiments in the brewery itself. Secondly, and more generally, pneumatic chemistry (that is, the chemistry of gases) had recently become a productive area of research, increasing in interest and significance through the work of Stephen Hales in the 1720s, Joseph Black in the 1750s and Henry Cavendish in the 1760s. This work confronted chemists with a proliferation of newly discovered substances whose properties, in isolation and in combination with liquid and solid substances, both demanded investigation and challenged many received ideas. Priestley's work was going to further this historical development, which would culminate not in Britain, but in France, with Lavoisier's work on oxygen, and the radical restructuring of the whole science of chemistry which followed from it.

After his brewery experiments, Priestley adopted the techniques of Joseph Black and Henry Cavendish for obtaining fixed and inflammable airs (carbon dioxide and hydrogen). His experimentation on fixed air was focused on its effects upon animal and vegetable life, and he noted in detail the differing times taken to kill mice, frogs and plants, as well as the effect of fixed air upon flower colours. He also examined processes which diminished the amount of air in which they took place, such as candle-burning, explaining this phenomenon by envisaging some portion of the atmospheric air to be somehow precipitated out of it during the process, and accounting for this by conceiving air as being composed to two substances, one being precipitated, the other left unaffected. The sequence continued with a great variety of experiments, many of them concerned with the noxious qualities of these gases, but also producing the discovery of marine acid air (hydrogen chloride) and using the calcination (oxidation) of metals to test Hales's view that common, or atmospheric air was absorbed when diminution of volume occurred. During calcination, Priestley showed, only one-fifth of the atmospheric air was consumed, and what remained was unable to support calcination and was noxious, confirming once again his two-part view of the composition of the atmosphere (what we now call oxygen and nitrogen). He also at this time (1772) isolated oxygen, extracting it from saltpetre, noting its property for supporting combustion, but regarding it only as a good sample of pure atmospheric air.

For this long and productive sequence of experimental enquiry and discovery, Priestley received the award of the Royal Society of London's prestigious Copley Medal. But he did not rest content with what he had so far achieved and continued pursuing the question of the composition of the atmosphere *via* the calcination of metals, particularly through the composition and decomposition of red lead (lead oxide). He believed first that the species of air involved was fixed air, with some reason, for lead oxide easily absorbs any fixed air present, which is then released by decomposition. The decisive step came when, in June 1774, he extended his range of calcined substances. When he decomposed calcined mercury (mercuric oxide), he obtained air which supported burning remarkably and was non-soluble in water. The same air seemed obtainable from nitre (potassium nitrate). As Priestley himself engagingly admitted, he had 'no idea at the time to what these remarkable facts would lead' and had simply mentioned his observations to scientific colleagues.

What has just been described is often regarded as Priestley's 'discovery of oxygen', but this is a less than helpful way to understand just what he was doing at this time. He was still examining the composition of atmospheric air and, experimentally convinced of atmospheric air's two-part nature, was especially concerned with that property of it which supported respiration. He did not, therefore, set out to 'discover oxygen', but to understand that fundamental support of human and animal life, the breathability of air. In so doing, he obtained a gas which, after Lavoisier, the world came to know as oxygen, but Priestley himself never accepted it as such, because the theoretical account of calcination which he accepted precluded any recognition of the substance and properties which Lavoisier named oxygen – a complex situation which will be examined shortly.

Priestley's immediate concern was to ensure that this potential new species of air was in fact distinguishable from others, such as nitrous air (nitric oxide), which he established by showing that unlike nitrous air, the new gas did not diminish atmospheric air, and neither did it extinguish a candle after standing over water, as nitrous air did. He began to realise its nature when he moved to breathability tests, finding that a mouse survived in the new gas more than twice as long as in common air. This confirmed his view that his discovery was akin to common air in its ability to support respiration. What was preventing an abandonment of his long-held persuasion that there was, in his own words, 'no air better than common air'? What was it that maintained this conviction so strongly and which for so long prevented him from a thorough recognition of his discovery? A likely suggestion is that it was extremely difficult for him to achieve the realisation that there could be a substance 'better' than the atmospheric air which God had made generally available throughout creation. What obvious purposes would such a substance fulfil? Why was it mixed with a non-supportive substance to form atmospheric air? The idea of such a substance was not one which could occur easily within Priestley's vision of nature.

Priestley continued his breathability tests and subjected the new gas to further tests of goodness, concluding that it was up to five times as good as atmospheric air. He eventually tried breathing it himself. 'Who can tell', he remarked, 'but that, in time, this pure air may become a fashionable article in luxury'; but he also warned against over-consumption of his powerful new gas.

He named it 'dephlogisticated air', a term requiring some explanation. It meant air that supported combustion, insofar as it readily absorbed 'phlogiston' from combustible matter; it was 'dephlogisticated' in the sense that it lacked phlogiston when in its pure state, but was highly capable of absorbing it. Phlogiston was the name used by eighteenth-century chemistry to designate a supposed principle of combustion. The more phlogiston a substance contained, the more combustible the substance was. Its use as an explanatory device was promoted in eighteenth-century chemistry by the German, Georg-Ernst Stahl, and by Priestley's time it was a widely used concept which had no serious rivals as an explanation for combustion. By this time, however, one notable quality had been added to phlogiston. Metals, on calcination, gain weight, yet the phlogistic theory indicated that during calcination metals were losing some of their substance namely phlogiston, the part of them which burned away. The solution to this quandary was to theorise that phlogiston was a substance possessing negative weight. Thus a substance was rendered *lighter* by its possession of phlogiston, *heavier* by its loss of phlogiston. This theoretical development was by no means as unacceptable as it may sound, for eighteenth-century science regularly employed notions of weightless substances. Both electricity and heat, for example, were widely regarded as highly tenuous and weightless material substances.

In thinking of phlogiston in this way, Priestley was simply doing the same as the vast majority of contemporary scientific opinion, and in continuing to maintain the phlogistic account of combustion he was also in good scientific company. Priestley, however, is singled out for special attention over the issue of phlogiston. This is not simply because he continued to support phlogiston after Lavoisier's destructive experimentation. Additionally and crucially, it was because Priestley discovered and distinguished the substance known as oxygen which Lavoisier in a very short time made the centrepiece of an anti-phlogistic chemistry. How, the question runs, having discovered oxygen, could Priestley have failed to agree with Lavoisier's oxygen theory of combustion, which saw combustion as being supported by atmospheric oxygen absorbed during burning, thus increasing weight? How could Priestley fail to see the obvious implications of his own discovery? Such questions are the wrong questions to ask, because they are not properly historical questions at all. In substance, what such questions reduce to is to ask: why was Priestley not Lavoisier? Put like that, the issue becomes obviously inapposite. Priestley's concerns were not by any means the same as Lavoisier's, although they overlapped. Priestley's primary concern was the goodness of air, and on that issue he made remarkable progress. Lavoisier's concerns were broader. For him, oxygen was part of a more general quest for a theory of the aeri-form state and a theory of acidity (the term oxygen means acid principle), as well as theories of combustion, calcination and respiration. It was the fate of Priestley and Lavoisier to converge upon the same newly-discovered gas from widely diverging practical and theoretical interests. A common interest in a new substance, even agreement on certain obviously observable properties of that substance, need not imply the same ways of visualising its significance, for its significance will tend to be functions of the particular and different scientific interests at stake. Ironically enough, Priestley was acutely aware, in general terms, of this quite regular feature of scientific life:

> . . . for we may take a maxim so strongly for granted, that the plainest evidence of sense will not entirely change, and often hardly modify our persuasions; and the more ingenious a man is, the more effectually he is entangled in his error; his ingenuity only helping him to deceive himself, by evading the force of truth.

Priestley was capable of mounting an ingenious defence of phlogiston, but this need not be seen as a signal theoretical failure negating a run of experimental success. His contemporaries did not view his achievement this way, and neither should we. It should also be argued that over-concentration on the issue of phlogiston distorts the historical interest and significance of Priestley's labours in science and intellectual life generally. By any standards, his career up until the phlogistic debate had been extraordinarily productive, and he still had many happy and productive years ahead of him, particularly at Birmingham, in the company of such men as Josiah Wedgwood, the industrialist, the doctor scientist Erasmus Darwin, and the engineer James Watt. They were all, with Priestley, fellow-members of the Lunar Society, a grouping of men of science and industry which has been termed the advance guard of the Industrial Revolution. They were indeed remarkable innovators in their respective fields, but Priestley was not destined to live out his final years in such select, congenial and stimulating company. As the tenor of political life grew increasingly repressive after 1791, Priestley found it advisable to quit Britain and move to America where his sons William, Henry and Joseph had already settled, and it was there that he passed the last ten years of his life.

The land of the free turned out to provide less of a haven of peace and stability than might have been expected. It cannot quite be said of Priestley, as has been said of Tom Paine, that having escaped the English treason laws, he was destroyed by the disapproval of his American neighbours, but Priestley undoubtedly encountered disapproval in America. Although America had come to provide a home for many religious dissenters, by no means all of them approved of Priestley's Unitarianism, which did not acknowledge the divinity of Christ, so his views were not universally welcomed. It was a comparable story in politics. Priestley, as ever, could not remain silent and was publicly critical of the Adams adminstration's increasing involvement in the politics of European war. American fears of sedition and a legislatively repressive response paralleled the course of events in Britain a few years earlier and culminated in pressure in the upper echelons of the administration to move against Priestley. Adams' response, though it ensured Priestley's safety (the two had been friends), was effectively humiliating: 'His influence is not an atom in the world'. Priestley was seen by Adams as something of a spent force, capable of creating minor public difficulty, but posing no major threat. His great reputation as a libertarian and friend of the French Revolution nevertheless proved an irresistible target to sections of the press, and he was attacked as a revolutionary, and a danger to true religion and law and order.

Yet if these negative themes of his life remained with him to the end, so too did the positive. Thomas Jefferson was a great admirer and elicited from Priestley plans for educational reform in Virginia. Priestley remained intellectually active to the end, but in America there was a sense of isolation from the centres of contemporary development in England and Europe. Priestley continued to write on religion and also on science, publishing in both Britain and America. He corresponded with Humphry Davy, pleased to have found a young ally who both advanced the chemistry of gases yet remained sceptical about French theory. Priestley retained his commitment to phlogiston unstintingly, arguing for it in the pages of the Philadelphian *Medical Repository*. Back in England, his most recent work was reviewed thus in 1802: 'It contains a number of interesting experiments, performed by this venerable philosopher, principally with a view to establishing his favourite hypothesis of phlogiston, to which he still adheres'.

'This venerable philosopher': the phrase, however honourably intended, has the sound of a politely uttered death-knell. Priestley was fighting a losing battle, long-past for most of his contemporaries. To be venerable, in this context, is once to have been great, to be revered for past achievement rather than present creativity. Thus venerated, Priestley died in February 1804. If his closing years in America have a faltering, fading air, then that is in major part because of the long maintained vigorous achievement which went before. In any case, the greatness of his life and work has always and rightly outlived his marginalisation in America. To remember Priestley as 'great' is to use an almost technical term, for, rarely among scientists, Priestley qualifies in general terms as one of England's great men, because his life bore many features which conveniently fit our conventional stereotypes of greatness in public life: discoverer, champion of liberty and free speech: honest, outspoken, determined, fearless, unwavering in commitment, but peace-loving withal. All true perhaps, yet there are now other important ways to remember Priestley, specifically with regard to science, and they merit concluding emphasis. At a time when science-induced change increasingly re-writes the plot of history as a bad science-fiction story – in a world, that is, of test-tube babies and Star Wars defence policy – Priestley's conception of the meaning of science is salutary. That science can and ought to be intimate with the pursuit of freedom and justice; that it is not, therefore, a neutral activity in isolated enclosure, but is instead an activity constantly possessing ethical and political meaning; such thoughts were second nature for Priestley and should be for us.

ANTOINE-LAURENT LAVOISIER:
The Chemical Revolution
Maurice Crosland

As science gathered momentum from the seventeenth century, it inevitably became enmeshed in politics. In France in particular, scientists became involved in the state, because the French Academy of Sciences was state-funded and provided salaries for its members. Eminent amongst these was Antoine-Laurent Lavoisier (1743–1794), who, from the 1770s, changed the face of chemistry. Lavoisier, however, was up to his chin in politics in another way, having a post in the tax-collecting department, and for this he lost his head in the Revolution.

More than two thousand years ago some primitive chemical technology was already in existence. This involved such practical processes as the extraction of metals and the production of ceramics and glass, dyestuffs and alcoholic liquors. The ancient Greeks contributed some influential theory, for example that all matter is essentially composed of a mixture of four so-called 'elements': earth, water, air and fire, where these terms are given the widest possible interpretation. (Thus any liquid would be described as 'water'.) The growth of alchemy, especially from the first century AD was encouraged by the belief that there is really only one basic matter, on which different 'forms' can be imposed. Although the quest for gold provided one strong motivation for undertaking primitive chemical experiments, and the later search for medical remedies provided a further utilitarian goal, by the seventeenth century the argument gained strength that the transformation of matter was worthy of study for its own sake. Prominent in this movement to study chemistry as a science was the Anglo-Irishman Robert Boyle (1627–91).

By the early eighteenth century, therefore, although chemistry had largely shaken off the legacy of alchemy, it was not yet clear in what direction it would develop. An unsuccessful attempt was made to give it the coherence of natural philosophy (or physics) by applying Newtonian principles to it. Basic techniques of analysis were slowly emerging, making it possible to distinguish, for example, between different alkalis, like soda and potash, and different 'earths' like lime and magnesia. At the same time the rate of discovery of new substances was increasing and a system of classification was necessary. Much of eighteenth-century chemistry could be considered as a branch of natural history. Lavoisier was to play down the natural history aspect, emphasise the logic of chemical composition and reaction, and give chemistry a new direction as a physical science.

The France into which Antoine-Laurent Lavoisier was born in 1743, was a focus of the Enlightenment, a period when many aspects of the authority of church and state came under attack from influential writers. Voltaire, after dabbling in science, had already turned his attention in other directions, but several years were to elapse before Rousseau would write his influential *Discourse on the Sciences and the Arts*. Though this was clearly a period of intellectual ferment, few could have foreseen anything so extreme as the political and social revolution which, in its most violent period, was to carry away many good Frenchmen, including Lavoisier himself.

In the eighteenth century France and Britain were the two great powers, Germany still being no more than a collection of semi-independent states whose greatest contributions to learning and to science lay in the future. By 1800 France had a population of approximately 27 million, compared to Britain's 11 million. The wide range of French traditional technology is illustrated in Diderot's great *Encyclopédie* (1751), but much industry was controlled by the state and lacked the flexibility, independence and initiative which were to make Britain the cradle of the first industrial revolution. France, however, already had a high reputation in the cultural and intellectual spheres. In the seventeenth century, academies had been founded concerned respectively with language and literature, the fine arts and science, and it was in some of these academies, rather than the universities of the period, that science and learning flourished. State patronage of science was much greater than in England and the mediocrity of the Royal Society of London in the eighteenth century compares unfavourably with the high standard of the Académie Royale des Sciences in Paris.

Paris was very much the city of Lavoisier. This was to be the scene not only of his birth and education but also of most of his work. The son of a lawyer, he was able to attend one of the best schools in France, the Collège des Quatre Nations, founded by

Cardinal Mazarin. There he received a sound classical and literary education and gained a prize in rhetoric. In the senior school Lavoisier was also able to study a little science, mainly mathematics and astronomy. To learn mineralogy and chemistry he had to wait until after leaving school, when he was able to take advantage of the popular lecture course of Guillaume-François Rouelle at the Jardin du Roi, where plants of medicinal value were cultivated and public lectures were given.

The first sciences to which Lavoisier made a contribution were mineralogy and geology. The geologist Guettard enrolled him as an assistant and they went on several expeditions in northern and eastern France. Lavoisier made a special study of gypsum, which is closely related to 'plaster of Paris' (calcium sulphate), and this was the subject of his first paper, presented in 1765 to the Academy of Sciences. Lavoisier continued to draw favourable attention to himself, and in 1768, at the early age of twenty-five, he was elected to the Academy in the most junior rank.

Although the Academy was to provide Lavoisier with the colleagues he needed to encourage and criticise his work, the stage on which to present it, and the principal journal in which to publish it, one must not think of membership of the Academy as employment. Only a few senior academicians received salaries. It provided honour and fame, for which the young Lavoisier thirsted, but for employment he would have to look elsewhere. After leaving school he had followed the family tradition and studied law. This was an appropriate training for someone who was to join a private consortium which collected taxes for the government. Lavoisier's paid employment, therefore, was not as a chemist but as an inspector of taxes. He was able to finance his experiments with the profits of the tax business. The expensive and elaborate apparatus which he had constructed to his specifications provided a contrast with the very basic equipment used by Priestley and, later, John Dalton.

Lavoisier was soon to marry the young daughter of another tax official, a girl with definite artistic and linguistic talents. She was to become the illustrator of her husband's textbook of chemistry and translator of key works in English, a language with which he was not familiar. She also acted sometimes as his amanuensis in the laboratory which he set up at the Paris Arsenal. She later described her husband's commitment to science, albeit on a part-time basis. He would get up early in order to devote two hours to science before going to his paid employment: he would regularly set aside a further three hours in the evening. But one day a week he would spend the entire day in the laboratory, surrounded by friends and associates. This, she reported, was his 'day of happiness'.

Lavoisier came to the Arsenal in 1776 during the brief ministry of Anne Robert Jacques Turgot. After the Seven Years War, in which France had lost India, French armies could no longer depend on the saltpetre (potassium nitrate) previously imported from that sub-continent, as the principal constituent of gunpowder. Turgot established a national gunpowder administration and appointed Lavoisier as one of its members. It was Lavoisier who first established the chemical composition of saltpetre. One might think of this applied research as being a diversion from Lavoisier's main scientific work, but we shall see that the chemist knew how to put all his experience to good use.

Lavoisier's switch of early interest from geology to chemistry came about through a decision in 1766 to study in turn each of the four Aristotelian elements: earth, water, fire and air. He began with studies of earth and water, but it was his studies of fire and particularly of 'air' which were to be crucial. Before the eighteenth century there was no understanding of the gaseous state. Any fumes or vapours would be described as 'air', supposedly a single substance, one on which Robert Boyle in the late seventeenth century had carried out a number of physical experiments. It was another British natural philosopher, the Rev. Stephen Hales, who asked in the *Vegetable Staticks* (1727):

> ... may we not with good reason adopt this now volatile *Proteus*[1] among the chemical principles ... notwithstanding it had hitherto been overlooked and rejected by chemists ...?

The Scottish chemist, Joseph Black, was to show in 1755 that, in addition to atmospheric air, there was a second kind of air ('fixed air') contained in substances like chalk. Black's discovery of carbon dioxide was followed in 1766 by the discovery of hydrogen ('inflammable air') by the wealthy recluse and natural philosopher Henry Cavendish. Joseph Priestley was subsequently to isolate many new 'airs' or gases, one of which was to provide the key to the chemical revolution. Priestley interpreted it in terms of the current theory of phlogiston but Lavoisier was later to call it oxygen.

The idea that many substances contained an inflammable prin-

---

[1] *Proteus* – a god in Greek and Roman mythology, fabled to assume different shapes.

ciple called 'phlogiston' was widely accepted by chemists in the mid-eighteenth century. In combustion, whether of wood, say, or metals, 'phlogiston' was supposed to be given off, leaving an ash or 'calx' behind. Readers who know that this calx was often what we should call an oxide might, nevertheless, agree that the theory was not as irrational as it may seem to the modern chemist. Most people who have not studied chemistry tend to think instinctively of something being *given off* in combustion, rather than something being *taken in*, as Lavoisier was later to claim. It did not seem to matter to eighteenth-century chemists that phlogiston could never be completely isolated. Yet the discovery of hydrogen gas gave a boost to theories of phlogiston since, as an inflammable gas which burned leaving no appreciable residue, its discovery appeared to give some experimental basis to a belief in phlogiston. It should be emphasised that there were many different theories of phlogiston, so that Lavoisier eventually had the task of combating not one precise and easily falsifiable theory but a whole family of theories, some very vague and infinitely adaptable.

The year 1772 has been described as 'the crucial year' in Lavoisier's scientific career, since it was then that he began a series of systematic experiments on combustion which were to reveal an anomaly in the current theory of phlogiston, thus precipitating a 'crisis', which was to lead eventually to a 'revolution' in chemistry. The French chemist found that when certain substances like sulphur or phosphorus are burned (or later when metals like tin and lead are strongly heated and change into an ash or calx), instead of there being a *decrease* in weight due to the escape of phlogiston, the product showed an *increase* in weight. Hitherto most chemists had shown little interest in quantitative considerations – chemistry was essentially a *qualitative* study. It was Lavoisier more than anyone else who transformed it into a *quantitative* science. Following Joseph Black, he brought the balance into the laboratory. It has even been said that he carried over into his chemistry some of the principles of the book-keeping involved in his tax affairs. He based his chemistry on the principle of conservation of matter – that there is an equivalence in weight (or mass) in every chemical reaction between the reactants and the products. This had been partly concealed previously by chemists not taking into account the escape of gases or (even more important) the absorption of gases, e.g. from the atmosphere.

We have emphasised that there were several different theories of phlogiston current in the 1770s and 1780s. Those chemists who took the increase in weight seriously tended to explain it by attributing negative weight to phlogiston. (This was not as irrational as it may sound.) One should not, therefore, assume uncritically that the demonstration of a gain in weight constituted a 'crucial experiment' which 'falsified' the theory of phlogiston. Lavoisier was to spend more than ten years carefully building up an alternative theory before, in 1785, he finally launched a frontal attack on phlogiston which, he said, was now completely supplanted by the oxygen theory. Phlogiston had become an *ad hoc* hypothesis, a vague principle which, because it could be adapted to explain everything, in fact explained nothing. The oxygen theory, on the other hand, was increasingly to become vindicated by its predictive value.

But to return to the early years: having demonstrated the gain in weight, Lavoisier's problem was to explain it. A century previously Robert Boyle had noted an increase in weight when lead or tin was heated in a sealed flask and air was then allowed to enter. He had attributed the increase to 'fire particles' passing through the glass. In 1774 Lavoisier was to repeat these experiments more carefully and attributed the gain in weight either to the (whole) atmosphere or to a constituent part of the atmosphere. This research, however, constituted only one of several strands which were finally to make up the fabric of a complete oxygen theory.

Lavoisier had meanwhile met Priestley in Paris, and Priestley had mentioned an 'air' he had prepared from red mercury calx. During the next year Lavoisier did some similar experiments but was drawn into a study of saltpetre for the gunpowder industry. He managed to improve the quality of French gunpowder, and consequently the range of French artillery, but we are here more concerned with a theoretical advance. He showed that saltpetre was a nitrate of potash, i.e. it could be made in the laboratory from nitric acid and potash, and concluded that:

> Nitric acid . . . is nothing other than nitrous air, combined with an approximately equal volume of the purest part of common air, and with a considerable quantity of water.

When Lavoisier wrote this in 1776, he had become more interested in the composition of acids than in the problem of combustion, and it is no surprise that he presented a general memoir on the nature of acids to the Academy in the following year. Intensely ambitious, Lavoisier thought he was in a position to put forward a comprehensive chemical theory of acidity. Generalising from a few examples such as nitric acid and sulphuric acid, he concluded that all acids contained a principle similar to the purest part of the air, which he called *oxigine* (later: *oxygène*), from the Greek, *oxus* = acid and *geinomai* = I engender. It was unfortunate that Lavoisier had not been able to decompose 'muriatic acid' (known to us as hydrochloric acid) or some other common acid which does not contain oxygen. This would have prevented him nailing his flag to the mast by the introduction of a term implying falsely that all acids contain oxygen, a term curiously still used in modern chemistry. Since the decline in the teaching of classical languages, however, it is only a minor embarrassment for English-speaking people, although the Germans still call oxygen *Sauerstoff*, or 'acid stuff'.

Yet Lavoisier is remembered less for the fact that he was 'wrong' about acids than that he was substantially 'right' about combustion. Even here the historian might not overlook that Lavoisier's oxygen gas was effectively a compound of 'oxygen principle' and 'caloric'; 'caloric' being the name Lavoisier gave to matter of heat which is given off in combustion. It was left to the next generation to show that heat is not a material substance. Lavoisier, despite the fact that he could not weigh caloric, insisted that it was a material substance because of the powerful effects it had on the senses. This rationalist chemist, who later boasted that he had effectively kicked out of the front door the 'metaphysical' phlogiston, was guilty of allowing the no less metaphysical caloric to creep in by the back door.

Overlooking these quibbles, however, we must give Lavoisier credit for having substituted a tangible material substance, oxygen gas, for the chameleon phlogiston. Overlooking too his early confusion of oxygen with 'fixed air' (carbon dioxide) and several other blind alleys and false trials, we may refer here to a classic experiment later described by Lavoisier. In a carefully controlled quantitative experiment he showed that, when a little metallic mercury was strongly heated for several days in a confined volume of atmospheric air, approximately one-fifth of the air was used up, leaving a gas ('azote', later called nitrogen) in which a taper would not burn. At the same time the surface of the mercury was covered with a red powder. When this powder (oxide of mercury) was heated, the reverse reaction took place:

$$\boxed{\text{red mercury calx}} \quad = \quad \boxed{\text{mercury} + \text{oxygen}}$$

Thus ordinary atmospheric air is essentially a mixture of two gases: nitrogen and oxygen, the latter allowing a candle to burn more brightly than in ordinary air.

If Lavoisier owed a little to Priestley for his isolation of a new air, he owed more to a second British man of science, Henry Cavendish, who had observed that when 'inflammable air' (hydrogen) was exploded with ordinary air, a small amount of dew was formed. Lavoisier learned of this in June 1783, when Cavendish's former assistant was on a visit to Paris. He immediately repeated the experiment, showing that under suitable conditions:

$$\boxed{\text{hydrogen} + \text{oxygen}} = \boxed{\text{water}}$$

He also carefully measured the proportions in which hydrogen and oxygen combined. Not satisfied with the complete logical rigour of this synthesis of water, Lavoisier went on to decompose water into its elements by allowing it to drip slowly down a gun barrel heated to red heat. The oxygen in the water combined with the iron barrel, leaving hydrogen gas, which was collected. This demonstration of the composition of water was important to Lavoisier in several ways, not least because it enabled him to explain satisfactorily the reduction of a metal calx (oxide) in hydrogen, an experiment which previously had only been explained plausibly by those like Priestley, who interpreted hydrogen to be a form of phlogiston. Lavoisier had a genius for devising quantitative experiments but he was quite capable of adjusting results which did not fit in with his ideas. A recent study of his laboratory notebooks has revealed that, when he set up an experiment which did not work very well, instead of re-designing the experiment he tended to salvage what he could and make additional assumptions in his calculations.

The elemental spring water of the poets had now been cruelly analysed and shown to be a compound of two gases. The very air necessary for human life was shown to be a mixture of another pair of gases. The common sense world was dissolving and being replaced by a rigorous new quantitative chemistry, largely based on an understanding of the role of gases in chemical reactions. One gas in particular, oxygen, was at the centre of the whole system. Combustion was seen as combination with oxygen. Oxygen could now be used to explain certain industrial processes which had developed by rule of thumb. Thus in the burning of sulphur to produce 'oil of vitriol' (sulphuric acid), the chamber door had traditionally been opened from time to time to 'sweeten the air'. This could now be interpreted simply as oxidation.

Respiration had always been something of a mystery. It could now be seen as a kind of combustion. In the seventeenth century certain philosophers, notably Descartes, had tried to reduce certain physiological actions to crude mechanics. Lavoisier was rather more successful in explaining respiration in purely chemical terms. He pointed to the analogy between the burning of a candle in a bell jar and the respiration of a guinea-pig:

(i)   both used up oxygen;
(ii) both produced 'carbonic acid gas' (carbon dioxide);
(iii) both left a residue of nitrogen;
(iv) both gave out heat (carefully measured with an ice calorimeter).

One outstanding difference was that in combustion a large amount of heat is produced in a short time, whereas in respiration heat is produced only slowly. Lavoisier, therefore, concluded in 1783 that respiration is a slow combustion. In the last years of his life he carried out further respiration experiments on himself and his assistant.

If Lavoisier's contribution to chemistry constituted a 'revolution', as is often claimed, it went beyond oxygen and chemical reactions. Indeed, it began with the very principles of chemical *composition*. Up to the eighteenth century chemical composition was usually explained in terms of the four so-called 'elements' of Aristotle, or the three 'principles' of Paracelsus, 'salt', 'sulphur', and 'mercury'. Robert Boyle in his *Sceptical Chymist* (1661) criticised such elements and principles, but was not able to put anything satisfactory in their place. One of Lavoisier's great achievements in his textbook of chemistry, published in the historic year 1789, was to draw up a list of thirty-three simple substances or elements which constituted the building blocks of the new chemistry. These included oxygen and hydrogen, phosphorus and carbon, copper and zinc, and even a few 'earthy substances' like lime, which resisted decomposition even on strong heating. Lavoisier characteristically gave a purely operational definition of an element as something which had not *so far* been decomposed into any simpler substances. This sensibly allowed for more powerful methods of decomposition which lay in the future, such as the electric current.

Chemistry up to this time had been dogged by trivial names, often inherited from the alchemical past. Name like 'oil of vitriol', 'butter of antimony', 'flowers of sulphur', and 'Glauber's salt' were an offence to the logical and systematic mind of Lavoisier.

Collaborating with three other leading French chemists, whom he had converted to the oxygen theory – Guyton de Morveau, Berthollet and Fourcroy – Lavoisier led a reform of chemical nomenclature on rational lines. Simple substances were henceforth to have simple names; compound substances should have compound names suggesting their constituents. In the new system, the very name *oxide of mercury* indicates a compound of oxygen and mercury. The French chemists introduced a system of suffixes to denote different proportions of the constituents. Thus there were sulph*ates*, sulph*ites*, and sulph*ides*, all compounds containing sulphur but with a decreasing proportion of oxygen. This was clearly a theory-laden nomenclature, which boldly assumed the validity of the new oxygen theory. Instead of proposing the use of the new terms from some future date, or calling an international conference, as would be done today, the French chemists decided that they would immediately put the new terms to use. In other words, after 1787 people could not understand the publications of the new French school of chemistry without having studied the new nomenclature. Pioneers like Priestley were forced to re-learn their chemistry and in the most humiliating way, i.e. assuming the principles of the oxygen theory, which Priestley consistently opposed.

In both the oxygen theory and the nomenclature Lavoisier was concerned above all with the chemistry of minerals, what is now called inorganic chemistry. Organic chemistry, which developed out of the study of animal and vegetable products, was then in its infancy and Lavoisier could do little more than begin the study of a few problems, notably alcoholic fermentation, and suggest how the proportions of the constituents of such compounds as alcohol and oils might be calculated after combustion. His chemistry was essentially the chemistry of elements rather than of atoms. Although several previous chemists had speculated about the atomic nature of matter, Lavoisier rejected atoms as metaphysical. It was left to the Manchester chemist, John Dalton in 1808 to publish a chemical atomic theory and assign relative weights to these atoms. Thus for Dalton each element is characterised by atoms of a specific weight or mass. In a sense, therefore, Dalton was able to complete the chemical revolution of Lavoisier based on elements. It should be noted, however, that no step such as that taken by Dalton could have been taken before Lavoisier had drawn up a list of elements. Dalton's chemical atoms were atoms of Lavoisier's elements.

Another achievement of Lavoisier was to explain the gaseous

state. The difference between solids, liquids and gases depends simply on the amount of 'caloric' present. Caloric is a material substance which insinuates itself between the particles of ordinary matter and, if present in sufficient quantity, produces a gas. Before the eighteenth century gases could not be accepted as material substances with a status comparable with that of liquids or solids. Lavoisier owed much to the eighteenth-century British school of pneumatic chemists: Hales, Black, Cavendish and Priestley, but only he was able to stand back from the miscellany of experimental details and provide a general philosophy of gases. Indeed, Fourcroy had considerable justification in speaking of Lavoisier's theory as the new theory of gases (*'la doctrine pneumatique'*) rather than the oxygen theory or the 'antiphlogistic' theory.

In politics Lavoisier, like many other French intellectuals of his age, was critical of the power and privileges of the nobility under the ancien régime. He favoured a moderate reform, and in his view of society science had an important part to play in solving economic, social and agricultural problems. He had an experimental farm and in 1790 turned his mind to problems of subsistence. He argued that the food consumed by human beings should be related to their physical work rather than to their economic circumstances. He took an active part in the early stages of the foundation of the metric system, in which the chaotic feudal system of weights and measures was to be replaced by a rational system based on 'natural' units. In the final years of the ancien régime Lavoisier had been a member of many commissions established by the government and by the Academy of Sciences, some concerned, for example, with the conditions of the prisons and hospitals of Paris. In one Academy commission he had criticised Jean-Paul Marat as a charlatan, thus earning the enmity of a powerful radical journalist in the later Revolution.

Yet, as the Revolution became increasingly violent, Lavoisier's greatest unpopularity came from his deep involvement in the tax Farm. When he was arrested in 1793, during the Terror, it was not as a chemist but as a tax official, and when he was sentenced to the guillotine in May 1794, he was executed as a lackey of the old regime rather than as a symbol of the new science. The wonder is that Lavoisier's exeptional scientific talents did not save him. But the Terror was a period of extreme measures, with France in a state of war with half of Europe. The Academy of Sciences had been closed down and the Jacobins in power did not consider the long-term consequences of their actions. The mathematician Lagrange said shortly after the execution of the chemist that it had taken only an instant to cut off his head but it might take a hundred years to produce a man of equal talent. Less than three months later the Terror came to an end and a more constructive period began which witnessed the foundation of several major scientific institutions. The scientific community – with some notable absences – was reconvened in the prestigious *Institut*.

Lavoisier had undoubtedly been the leading light among the Parisian scientists of the 1780s. His association with three other prominent chemists in the reform of chemical nomenclature led to the description of the oxygen theory as 'the theory of the French chemists'. Lavoisier, however, angrily protested that it was not a communal theory – 'elle est la mienne' ('it is mine'), he said. His writing of a textbook, the *Traité élémentaire de Chimie* in 1789 can be seen not only as an attempt to codify the new chemistry, but also as a bid to present his personal view of chemistry to young people not committed to a theory of phlogiston. He never held a teaching position, yet through this book he was to teach the new chemistry to the next generation. Young Humphry Davy and Gay-Lussac both studied the *Traité*, and were to develop different aspects of the work of the great French innovator.

In 1789 Lavoisier helped to found a key scientific journal, almost the first specialist chemical journal, the *Annales de Chimie*. The editorial board included his colleagues involved in the reform of the nomenclature, and they continued publication after his death with only a brief interruption. That is not to say that they all shared identical views of the new chemistry. Fourcroy, for example, stressed the utilitarian aspects of chemistry and refused to sever the old links of chemistry with pharmacy. Berthollet collaborated with the great mathematical physicist Laplace to found a school for neo-Newtonian physics and chemistry at Arcueil, just outside Paris. Prominent younger members of this group included Gay-Lussac and Arago, both graduates of the post-Revolutionary Ecole Polytechnique. Gay-Lussac modified Lavoisier's theory of acidity and proposed a method of the analysis of organic compounds which was to be further developed in turn by his own student, Justus von Liebig (1803–73). Liebig was to become the founder of an important new German university-based research school.

Resistance to the new oxygen theory in the German states was strengthened by the emergence of nationalistic feelings, perhaps not unrelated to the fact that the phlogiston theory of Stahl could be seen as 'German'. In Britain the Revolutionary and Napoleonic

Wars had little effect on the gradual acceptance of the new theory. They did, however, intensify scientific rivalry, as exemplified in the parallel research of Humphry Davy in London and Gay-Lussac in Paris. Davy, in his isolation of new elements and his work on chlorine, gave a new look to Lavoisier's chemistry, whilst the influential Swedish chemist, Berzelius, reorganised the science around the principles of the new electrochemistry rather than around oxygen.

Lavoisier's work inspired a hunt for new elements and the original thirty soon became forty, then fifty . . . Some systemisation was imperative if chemistry textbooks were not to revert to being collections of miscellaneous recipes. The major advance in classification came in 1869 with Mendeleev's Periodic Table. Yet, although the name of Lavoisier will always be associated with elements, the most important contribution of the French chemist was simply to put chemistry on the map. Indeed, chemistry became the model science of the early nineteenth century, having thrown off for ever the dubious links with fraudulent alchemy. Chemistry was now showing itself to be a science not only of theoretical power but also of great practical utility and this was an important factor in the professionalisation of chemistry in the nineteenth century, before many other branches of science.

The tragic death of Lavoisier and the happy survival of his colleagues proved something of an embarrassment to them. His widow worked hard to keep his name alive, but proper tribute to the founder of the new chemistry had to wait for Jean-Baptiste Dumas in the mid-nineteenth century. And the Academy of Sciences, which spoke his name with pride, nevertheless delayed until the centenary of the French Revolution to pronounce the customary eulogy. The secretary of the Academy seized on the concept of revolution in order to explain the significance of Lavoisier's work. But in doing so, he was only confirming the prediction of his eighteenth-century compatriot. As early as 1773 the optimistic young Lavoisier had written that he expected his experiments to bring about a revolution in chemistry. Time has vindicated this prophecy, although the nineteenth century was to see the greatest advance in organic chemistry, the branch which in Lavoisier's time was least developed. Lavoisier was a believer in the Enlightenment concept of progress, but he could hardly have foreseen the enormous possibilities of synthetic organic chemistry, including an immense range of dyes which began to transform fabrics and fashions from the mid-nineteenth century onwards, and a powerful battery of drugs which has transformed the face of medicine in the twentieth century.

James Watt (1736–1819)   Michael Faraday (1791–1867)

Charles Darwin (1809–1882)

Louis Pasteur (1822–1895)

JAMES WATT:
Cross-fertilisation between Science and Industry
Donald Cardwell

How much have practical skills and industrial developments contributed to science? How far, in turn, are industrial and technological advance endebted to science? There are no simple and universal answers to these questions. There is little doubt, however, that close and fruitful links were forged between men of business and men of science during the years of the Industrial Revolution (from about 1760 to about 1850). James Watt (1736–1819) is an important representative of these developments. Deriving his science from the leading Scottish professors of physics and chemistry, he showed immense practical ingenuity in the major improvements he made in steam-engine design. He then went into industrial partnership with one of the leading iron-masters of the day. Watt's engine, in its turn, helped to focus the attention of scientists concerned with the nature of heat and its relations to motion and mechanical action.

James Watt was the first folk hero of the steam engine. As is usual with folk heroes various legends are associated with his name, the most familiar being that of the boy who sat watching a kettle boil and the one about the young man who, on being told that there was much heat in steam, went on to invent his engine. There is some truth in these legends, but there is more that is misleading.

Watt was born in Greenock on the Firth of Clyde, the son of a town councillor and merchant whose business included ship's chandlery and marine engineering. His paternal grandfather had been a teacher of navigational mathematics while his father and uncle were practised surveyors. On his mother's side he was related to George Muirhead, a professor at Glasgow University. It is hardly surprising that, as a boy, he was interested in mathematics and Newtonian mechanics. And it seems natural that he should have chosen instrument-making as a career, his training being first in London and then in Glasgow. In those days apprenticeship to an instrument-maker was one route into a career in science or technology. John Smeaton, the leading civil engineer of the century, and (Sir) William Herschel, the leading astronomer had both taken this route.

Through his kinsman, George Muirhead, young Watt gained the patronage of Robert Dick, Professor of Natural Philosophy, in Glasgow, and, with Dick's support, was appointed instrument-maker to the University. In those days Glasgow was a small town and its university – 'Glasgow College' – was correspondingly modest, boasting about a dozen professors with Medicine the largest faculty, with three professors. One of these, Joseph Black, the Professor of Medicine, became a good friend of Watt, as did the student John Robison, who was later to become Lecturer of Chemistry in the University. Robison and Watt carried out experiments on the pressure of steam to see if it could be used to drive machinery. They concluded it would be far too dangerous.

When Dick retired he was replaced by John Anderson, an energetic, colourful and rather abrasive character who later resigned to found his own university, the 'Andersonian', now Strathclyde University. But before he resigned he performed, if unwittingly, a hardly less valuable service. He had a model Newcomen engine, supposedly a small replica of a large, working engine, that he used to demonstrate its principles to his class in Natural Philosophy. The model had been broken and Watt was asked to repair it.

To understand what followed we must know something about the contemporary steam engine. The discovery of atmospheric pressure (14.7 lb per square inch, or 10.34 tonnes per square metre) and the correlative discovery of the vacuum in the seventeenth century had led men to try, without much success, to turn the new force to useful account. By 1712 Thomas Newcomen had succeeded with his mine-pumping engine. This consisted of a boiler with a vertical metal cylinder mounted on top and fitted with a piston. The piston was connected by means of a rod and chain to one end of a massive, pivoted timber beam the other end of which was linked to a pump rod. Steam from the boiler flooded into the cylinder so that the piston rose up and the beam tilted over. When the piston reached the top the steam was turned off and, simultaneously, a spray of cold water was injected into the cylinder. This condensed the steam, leaving a vacuum, so that the outside atmospheric pressure could drive the piston down again, the beam tilted back and the pump rod was raised. When the piston reached the bottom of the cylinder the spray was turned off, the steam was turned on and the whole cycle repeated. From the very beginning, in 1712, the engine was self-acting, that is to say the valves were opened and closed at the appropriate times by the movement of the great beam.

Two modifications were needed to make the engine work satisfactorily. The air that accompanied the steam from the boiling water had to be expelled; otherwise it would steadily accumulate until the cylinder was full of air and the engine would stop, 'air logged'. In the Newcomen engine the inrush of steam every cycle flushed the air out through the 'snifting' valve. The other requirement was that the cylinder should not be cooled down too much. Experience taught the old enginemen that some steam should be left in the cylinder. This meant that there was some back pressure to oppose atmospheric pressure but, at the same time, the cylinder remained warm and much less steam was required next time to heat it up. Accordingly the engine worked much more rapidly and was more economical in steam consumption.

Although atmospheric pressure provided the force for the working stroke, the actual agent was the steam pressure that pushed the atmosphere back, as against a spring. The engine was usually called a 'fire engine', although it was not understood at that time that heat energy was the ultimate agent. It was very well adapted for its purpose. It was powerful, reliable, profitable, safe and within the capacities of skilled craftsmen – millwrights, blacksmiths and the like – to build and operate. During the eighteenth century many 'fire engines' were put up in the Midlands, the North West, the lowlands of Scotland and, more particularly, in Cornwall and the North-East coalfield.

For all its revolutionary importance there was little public interest in the Newcomen engine. It is easy to understand why. The engine was largely confined to remote mining areas. And even in such places all that was to be seen was an engine house with a smoking chimney while the end of a wooden beam, sticking out through one wall, moved ponderously up and down.

Watt, however, was hardly likely to have been interested in the diffusion of knowledge about the 'fire engine'. He was an instrument-maker with a job to do. He mended the little engine and then ran it to see that all was in order. He found that the little boiler was unable to supply enough steam to keep it running for long. This puzzled him. The large engine did not run out of steam; why should the model? At this point he transcended the office of instrument-maker and became a natural philosopher. The little cylinder was hot when full of steam and relatively cold when the steam condensed. The same sequence took place in the large engine – but there was a substantial difference in scale. Galileo had discussed the scale effect in his *Two New Sciences*. More particularly Newton, in his *Opticks* (1704), had pointed out that a small body cools down more rapidly than a large one; the reason being that the ratio of the surface area, across which the cooling takes place, to the volume is much greater in the case of a small body than in the case of a large one. The phenomenon and its explanation were also discussed by Hermann Boerhaave, the influential chemist and teacher of medicine at Leiden University in Holland, in his *Elements of Chemistry*. The steam in the little cylinder would, accordingly, lose heat and condense more rapidly than the steam in the large cylinder. Thanks to his readings in contemporary science and talks with university friends the solution to the problem must have been relatively simple for Watt. But we should note here that thinking and arguing in terms of proportions was characteristic of eighteenth-century science.

Everyone recognised that the alternate heating and cooling of the cylinder was, to a degree, wasteful and while most would have regarded it as unavoidable – in the nature of the beast – some had speculated on how to reduce it. James Brindley, the self-taught English canal engineer, had tried using wooden cylinders, but his experiments were unsuccessful. Watt's approach on the other hand, was systematic: he quantified the problem. How much steam was used to work the engine? Curiously, no one had measured this before. And if steam consumption was excessive how could it be reduced? By careful measurements on the little engine he found that the boiler was supplying several times more steam than would be required just to fill the cylinder. Clearly there was a major dis-economy; the loss of heat through the cylinder walls was excessive.

The next step was to find a material that would absorb less heat than iron or brass. This means that he was thinking in terms of quantities of heat; the 'quantity of heat' being the measure used by Joseph Black who had just discovered the concept of 'specific heat capacity', or the quantity of heat needed to raise the temperature of a specific substance one degree Fahrenheit. What he was looking for was a material that would absorb heat less readily than any metal so that the walls of the cylinder would remain hot and excessive condensation of steam would not take place. His experiments convinced him that wood was the best material so he made a wooden cylinder, soaked it in linseed oil and baked it dry. The wooden cylinder filled with steam much more readily than the metal one. An economy had been made. But when he set out to find exactly the right amount of cold water to condense the steam he was surprised to find that a disproportionate amount of water was needed. In fact, water converted into steam can heat about six times its own weight of well-water to 212 degrees Fahrenheit; Watt had come across the phenomenon of latent heat of vaporisation. Black had already discovered this and was able to explain its general significance in the theory of heat. It was not important in the operation of the steam engine but it was the source of a legend, according to which Black instructed Watt about latent heat and thus prompted him to make a radical improvement in the steam engine. But the legend is inconsistent with the facts, for it was the heat lost through the metal cylinder that was the cause of the waste; the latent heat of steam was irrelevant. The loss would have been the same whether the latent heat was far greater or far less than it was. Watt himself explicitly denied the story which began with an account written by Robison some forty years later. However, the story fits nicely with the picture of the gentleman of science instructing the deserving artisan in the latest scientific discovery and then telling the good fellow to go off and put it to some practical use. Much the same story has been told about other notable inventions. It is a disturbing possibility that elements of this view of the relative functions of 'pure' science and technology may still influence thinking about their relative roles today.

The wooden cylinder was a failure. The economy had been purchased at the cost of reduced power. And the reason was soon clear. Injecting only enough cold water to condense the steam meant that the condensate – the mixture of condensing water and condensed steam – came out of the cylinder boiling hot. Now Black's immediate predecessor in the Glasgow Chair, William Cullen, had shown that tepid water will boil in a vacuum and will go on boiling down to below 100° Fahrenheit. But boiling water generates steam and steam in the cylinder must exert a pressure on the piston that will oppose the downward pressure of the atmosphere. The only way to prevent this would be to cool the cylinder and its contents down to below 100 degrees Fahrenheit; but this would reduce the economy of the engine. It might be thought that Watt's systematic procedure had taken him no further than Brindley's intuition had taken him. But this was not the case. Watt had clarified the problem, brought it into sharp focus, and so indicated the essential requirements. They were, that, for economy the cylinder must be kept hot all the time, for power it must be cooled right down once every cycle.

After much thought Watt was able to reconcile these apparently mutually exclusive requirements. The solution was to use two cylinders; one, in which the piston moved, kept hot all the time, the other, the condensing cylinder, kept cold all the time. To condense the steam, a valve in the pipe joining the two cylinders was opened whereat the steam rushed into the cold cylinder to be condensed in turn; and this went on until all the steam had been condensed and there was a near perfect vacuum in both cylinders. The process was almost instantaneous. When the piston reached the bottom of the cylinder, the valve would close, the piston would rise to the top and the cycle would be repeated. To this comprehensive solution, reached in early 1765, Watt added two refinements. He saw that it would be wasteful to allow cold air to drive the piston down. He therefore closed the top end of the cylinder, took the piston rod out through a stuffing box, and relied on hot steam to do the job that cold air did for the Newcomen engine. He now had a complete steam engine even though the steam was at a pressure not much above atmospheric. The remaining problem was how to get rid of the air that accompanied the steam from the boiler. He could hardly flush it out through the cold condenser. The best solution, he saw, was to fit an air pump to the condenser so that the air, together with the condensed steam, could be pumped out as the great beam moved slowly up and down. In this way, after much effort and thought, James Watt invented the efficient condensing steam engine that he patented, together with other ideas including a suggestion for an engine to produce rotative motion, in January 1769. Six months later Richard Arkwright patented his water-frame for cotton spinning and these two events can be taken as ushering in the industrial revolution.

Watt was far from typical of those engaged in the steam engine business. He came into it almost accidentally. This meant that he avoided the assumptions – and perhaps the prejudices – of practical engineers. His contacts had been academic. And this was significant. The Scots, unlike the English, understood and valued education. During the eighteenth century the Scottish universities flourished. Men who had studied medicine under Boerhaave – and their students after them – were most successful in raising the prestige of the medical faculties. Following Boerhaave's precept, academically inclined medical men studied chemistry and this included the study of heat. William Cullen and Joseph Black were by no means the only ones to do this. Watt was therefore unique among the early power engineers in having had direct and indirect contact with men who made scientific studies of heat. And while it would be wrong to assume that Watt was a mere disciple of

Black, it would be equally erroneous to ignore the ideas, the science, of such men. Watt cannot have avoided their influence. On the other hand, John Smeaton, the distinguished engineer whose background closely resembled that of Watt, had had little or no contact with chemistry and the study of heat. He was a practical man, albeit with a liberal and informed mind. When he first saw Watt's engine he recognised at once that it was superior to the Newcomen engine but argued that it was far too complex to be practicable. He was almost right. It was a laboratory invention; not really suitable for the rough and tumble of eighteenth century British industry.

Watt was ever the perfectionist. He fitted his early engines with a 'steam jacket' to reduce heat losses. This was a wooden cylinder lined with copper that enclosed the working cylinder. The space between the cylinders was filled with steam. Again, he began with a 'surface' condenser in which steam was condensed by contact with metal kept cool by cold water on the other side. But these luxuries proved too expensive. The steam jacket was abandoned and spray condensation reverted to in the condenser. The sacrifice of efficiency was small compared with the overall gain due to the separate condenser. His striving for perfection was also manifest in his invention of expansive operation.[1] Watt saw that it was wasteful to let a cylinder full of steam at atmospheric pressure *rush* into the condenser at zero pressure. Much better to turn off the steam before the piston had gone far down the cylinder and let the innate pressure of the steam complete the job. The steam in the cylinder, now isolated from the boiler, would continue to push down the piston, on the other side of which there was a vacuum, an empty void. The volume occupied by the steam necessarily increased as it drove down the piston and, just as necessarily, its pressure fell. In principle the pressure should fall almost to zero at the end of the stroke, but in practice this was unattainable. The aim was that every last drop of effort – or energy, in modern terms – should be squeezed out of the steam, every last bawbee on the fuel bill, made to do its duty. And what more could a good Scotsman want?

Quite possibly Watt got the idea of expansive operation by reading Smeaton's classic paper on water power (1759). Smeaton showed that if the water was to give up the maximum amount of energy to the water-wheel it should leave the mill with the minimum velocity. It should not *rush* away down stream.

---

[1] First described in a letter to William Small in 1769; patented in 1782.

To exploit the new engine Watt went into partnership with Dr Roebuck, a medical man and former student of Boerhaave's, but then running the Carron ironworks and some coal mines. The development of the engine proved slow and difficult. Roebuck's bankruptcy (1773) would have been a disaster had not Matthew Boulton stepped in and bought his share of Watt's patent. A Birmingham manufacturer, Boulton was the ideal partner for Watt. He had manufacturing resources, business contacts and a character that seemingly complemented that of Watt. While Watt was suspicious of others (sometimes with reason) and, on occasion, intemperate in his judgement, Boulton had an equable disposition. And Boulton, more than anyone else, more even than Watt himself, foresaw the world-wide potential of the new steam engine. As Boulton was later to remark to Boswell, 'I sell here what the world wants: power'.

Initially the partners concentrated on selling the engine to the mines in Cornwall, where fuel economy was of prime importance. Cornwall gave them useful experience and publicity but collecting royalties proved difficult and they often had to go to law. Piracy was easy, for an engine house was a private place and any able mechanic could surreptitiously fit a condenser to a Newcomen engine. Boulton saw that the market had to be widened; a rotative engine must be developed, one that could drive corn mills and textile machines. Accordingly, in 1782 and 1784 Watt obtained necessary patents for a rotative engine. One of these was for the 'sun-and-planet' gear, a clever device to circumvent a patent that Watt believed had been taken out on the application of the crank to the (Newcomen) steam engine. This showed, once again, that he could solve an apparently insoluble problem. And it confirms that he was an inventor of genius as well as a pioneer applied scientist. He was, in short, a one-man research and development department as well as a skilled service engineer. Furthermore, he and Boulton trained up a new generation of engineers. Aware of the national importance of all this work, in 1785 Parliament took the unprecedented step of extending Watt's patent monopolies for a further fifteen years.

In 1796 the partners opened the Soho Foundry in Birmingham for the manufacture of complete steam engines. Previously they had relied on outsiders for items like boilers and cylinders but the arrangements were unsatisfactory and open to abuse. The Soho Foundry has been called the first modern engineering works. What is indisputable is that by the end of the century the steam engine had changed from being a piece of mining machinery into a universal source of power. In 1819 Watt died, a wealthy man with a country estate and the satisfaction of knowing that his son and Boulton's son were running the firm of Boulton and Watt. He had turned out to be a good businessman as well as a prolific innovator, although not all his inventions were successful.[2] The only things Watt disapproved of were the increasingly popular high pressure steam engines; he thought they were far too dangerous. But high pressure steam enabled much more compact, powerful and, as it turned out, efficient engines to be built; engines that could be put on wheels, to run on tracks. And, ironically, the higher the steam pressure the greater the benefit from expansive operation. However, as Dr R. L. Hills has pointed out, the steam engine was thought of as a *pressure* engine. The engineers who followed Watt had little or no contact with the science of heat and therefore tended to ignore the role of heat in the operation of the engine. They knew that leaks of steam had to be prevented, that furnaces should be well designed, that friction must be kept to a minimum. But 'heat' was another and abstract matter.

One man who clearly recognised the importance of heat in the operation of the steam engine was the Frenchman, Sadi Carnot. In 1824 he published a comprehensive theory of the heat engine. He saw that the high pressure engine was more efficient than the low pressure engine because high pressure steam is also high temperature steam and it is the temperature difference that determines efficiency. He postulated Watt's expansive operation as one of the two essential features of the perfect, or 'Carnot' cycle. But his persuasive argument was ignored. Interestingly enough, in 1828 Samuel Grose, a Cornish engineer, discovered that if he 'clothed' (that is, lagged) the cylinder and pipes of his engine there was a great improvement in performance. But nobody seemed very interested.

The man who had the credit for establishing Carnot's theory was young William Thomson (later Lord Kelvin). In 1849 he gave an account of Carnot's theory and used it to explain the high efficiency of the high pressure – and therefore high temperature – Cornish engines. In this account the word 'thermodynamics' appeared for the first time. There was, however, a flaw in Carnot's and Thomson's reasoning. They made the common assumption that heat is conserved, that it flows through the engine like water through a water mill. Many people, indeed, thought of heat as a

[2] For example, the 'steam wheel' of 1769 was an early attempt to make a rotative steam engine, but it was never a success.

special sort of substance called 'caloric'. In fact, heat engines depend on the flow of some heat and the actual conversion of the rest into mechanical energy. A different concept of heat had already been established by James Prescott Joule, a young Manchester man who, as a boy, had gone with many others to watch the locomotives on the revolutionary new Liverpool and Manchester Railway. He was able to convert Thomson to the dynamical, or energy, concept of heat. This made it possible for the Carnot theory to be reformulated and the science of heat to be established on the basis of energy. And in this way it became a part of a new science with an old name: physics.

Watt's achievement was one of the foundation stones of the new science. It is a curious fact that Joule began his researches with an attempt to perfect a new power source, the electric motor driven by a battery. After much work he concluded, sadly, that such a combination could not compete with a good Cornish engine. Joule was a shy man who gave very few public lectures, and then only to informal groups. But in 1865 he made an exception. He travelled up to Greenock to read a paper in commemoration of the centenary of Watt's invention. Today the joule is the unit of energy, the watt is the unit of power. And this is appropriate for the great doctrine of energy and the science of thermodynamics are deeply indebted to the long research for power. One other achievement may, perhaps, be credited to Watt. Even the most patriotic Scot would find it hard to discover a Scottish name among those of famous engineers before Watt's time. After Watt Scottish names abound. Of course, engineering skills, like medical skills, were acceptable and valuable qualifications everywhere in Britain; everywhere, indeed, in the world. And there may well have been other reasons to account for the rise of the Scottish engineer. But the influence and example of James Watt should never be underestimated.

In the meantime the second folk hero of the steam engine, George Stephenson, the railway engineer, had made a wondering and appreciative public fully aware of the steam engine and what it could do for all people. By the end of the century the steam engine – locomotive, stationary and marine – had transformed life in this country and in most of the world. By then Thomas Newcomen was almost forgotten. Today he is remembered only by specialist historians.

MICHAEL FARADAY:
The Use of Pure Science
David Knight

Just 200 years ago almost everyone who pursued science did so as an amateur. Most scientists were therefore gentlemen of means. The founding of the Royal Institution in 1799 made a slight but important change in this situation by creating a small number of full-time posts for scientists. In its first half century, the Institution's two most distinguished scientists – Humphry Davy (1778–1829) and Michael Faraday (1791–1867) – were both poor boys who made good. Both excelled in experimental investigation and, perhaps because neither had a formal university training, they tended to neglect the mathematical dimensions of physics; a fact which seems to have acted in their favour. From the time of the Greeks, investigation of Nature had commonly been split into relatively separate theoretical, experimental and mathematical approaches, and the sometimes harmful but often fruitful tensions between them remained powerful in the nineteenth century.

We live in a world which depends on electricity. The first industrial revolution was based on steam engines, but the current one is founded on electronics. The pioneers of steam were practical men, and the science of thermodynamics came a hundred years after the first working engines. Theoretical analysis of actual machines led to fundamental understanding of heat, work and changes of state. While the steam engine thus gave more to science than it took from it, the electrical industry arose from scientific discoveries which were, at first, of no apparent use. This is the pattern which we have now learned to expect; but in the nineteenth century it was completely new. Francis Bacon, two hundred years before, had argued that experiments of fruit would follow from experiments of light; but this notion remained a hope rather than a reality. To Michael Faraday more than to any other man we owe the fundamental research which underlies the modern electrical industry: although when he died in 1867 the world was still lit by gas and oil, transported by horses and steam, and dependent chiefly upon paper for records and communications

Faraday was born on 22 September 1791, on the outskirts o London. His father was a blacksmith, and a member of a smal fundamentalist sect, the Sandemanians. Throughout the difficul years of the French wars, which went on right through Faraday's childhood and youth until the Battle of Waterloo in 1815, his family remained among the 'respectable' poor who had never needed parish relief; but his education was minimal. It is hard now to imagine becoming a great scientist without secondary schooling: what Faraday lacked was a training in mathematics, but there can be advantages in not being subjected to formal and dogmatic teaching.

In 1805 Faraday was apprenticed to George Ribau, a London bookbinder. Books then were expensive, and came out in paper wrappers ready to be bound in leather to the buyer's requirements, so bookbinding was a skilled and useful craft. In 1812 he finished his apprenticeship; by then he had taken advantage of his job to read some of the books he was binding and also joined a group of other young men anxious to improve themselves. This was an exciting period in the sciences, and especially in chemistry, revolutionised by Antoine Lavoisier at about the time Faraday was born. There were not yet 'two cultures', no gulf separated sciences and humanities, and anybody with intellectual interests was expected to know something of what was going on in science. In Britain, eminent chemists such as Humphry Davy, William Hyde Wollaston, and John Dalton were proving that chemistry was not just a French science, thus fighting the French wars in the intellectual realm. Faraday and his friends held discussions and debates, and did some experiments; his textbook was Jane Marcet's *Conversations on Chemistry* of 1806. In this classic work written for girls, the formidably bright Emily and Caroline learn the latest science (picked up by the author from Davy's lectures) with their tutor Mrs B. through discussion and experiment. Faraday decided that rather than pursue trade, he wished to be a natural philosopher, as men of science were then called.

The problem was that there were no careers open to such people. Society then worked on patronage – it mattered whom rather than what one knew – and Faraday wrote a letter to Sir Joseph Banks, President of the Royal Society and companion of Captain Cook, asking for his help. Banks was helpful to several young men, but did nothing for Faraday. Patronage is a hit-or-miss business. In 1812 a customer at the bookshop gave him tickets for Davy's lectures at the Royal Institution: Faraday went, took notes, wrote up a fair copy of them, bound it, and presented it with a note to Davy. Davy saw him, and advised him to stick to bookbinding, science being 'a harsh mistress in the pecuniary point of view'. Faraday was now working for another bookbinder, De la Roche, who had said: 'I have no child, and if you will stay with me you shall have all I have when I am gone'; but Faraday was sure his future lay with science. In March 1813 the laboratory assistant at the Royal Institution struck an instrument-maker, and was dismissed: the job was offered to Faraday. He accepted, and joined the laboratory where he was to spend the rest of his working life.

The Royal Institution had been started in the last years of the eighteenth century as a place where practical science could be taught to landowners and mechanics; but the social distance between these groups was too great, and the place never became a technical college. Instead, lectures were given to large and fashionable audiences, while research went on in the basement laboratories. Some of this research was in what we would call 'pure' science, and some in 'applied'; this distinction was not made in Regency Britain, where the contrast was between 'science' (systematic inquiry) and 'practice' (rule of thumb). Audiences got some lectures on agricultural chemistry, and others on recent discoveries as yet of no utility; because large numbers of women came, lectures were rhetorical rather than dry and formal, and were well illustrated with demonstration experiments. The value and the excitement of science were daily proclaimed in the Royal Institution, and there were also, sometimes, practical tips for the landowner or manufacturer. The message got through, and the place became one of the intellectual centres of the day. Davy was the person largely responsible for this. His father had been a wood-carver in Cornwall, and like Faraday he had had little schooling, being apprenticed to an apothecary. James Watt's son Gregory had boarded with the Davy family when sent to Cornwall in the hope that its mild winters would cure his tuberculosis; and Watt and his circle had become patrons of the young Humphry, getting him a post in Clifton as a chemical assistant in a clinic where gases like oxygen were being given to sufferers from chest complaints. Here he discovered the properties of laughing gas, and made a thorough study of all the oxides of nitrogen; and became friends with S. T. Coleridge and William Wordsworth. He was duly invited in 1801, at the age of twenty-two, to the Royal Institution as lecturer in chemistry. A small dark man from the Celtic Fringe, he had a great gift for language and a glittering eye to hold an audience, and proved an enormous success; he was soon promoted to professor.

He was also outstanding as an experimenter. Early work on the chemistry of tanning earned him the Royal Society's Copley Medal; and then from 1806 he turned his attention to electrochemistry. Volta had in 1799 invented the wet battery; and Davy was one of those who believed that, in it, chemical affinity was transformed into electricity. He thought that chemical affinity and electricity were manifestations of one underlying force. Applying an electric current might therefore break up otherwise stable compounds; and Davy tried it on the alkalies. He isolated the new metals potassium and sodium, which were lighter than water and had spectacular properties which made splendid lecture demonstrations; Davy made a clay volcano with potassium in it; poured on water, and watched it erupt. He hoped that electrical measurement of affinities might make chemistry into a fully quantitative science.

At this time there was no unified science of physics – optics, mechanics and electricity were all quite distinct – and chemistry seemed the fundamental science, dealing with the hidden forces of nature. For Davy, it was the disposition of electrical forces which determined chemical properties, rather than the presence or absence of components; he refused to believe with the French that oxygen was the source of all acidity, and went on to prove experimentally that there was none in the strong acid from sea salt, our hydrochloric acid. The greenish gas hitherto called 'oxymuriatic acid' became for Davy an element, chlorine. It seemed as though in his hands the whole science was once again in a revolutionary state; and the French Academy of Sciences awarded him a prize for his electrochemical researches, in spite of the war, with an invitation to go to Paris and collect it. In 1812 Davy was knighted – it was exceedingly rare to be thus honoured for science – and married a wealthy widow of intellectual tastes, Jane Apreece. It was to his triumphant course of lectures on chlorine that the young Faraday went; although only twelve years younger than Davy, Faraday never knew him except as Sir Humphry, everywhere recognised as the most brilliant man of science of the day.

Davy was not a systematic thinker or a methodical experimenter. His contemporaries, Dalton in Manchester, Berzelius in Sweden, and Gay-Lussac in France, sought falsifiable laws where Davy loved grand generalisations. His laboratory notes were kept on odd bits of paper with a scruffy quill pen, and one of Faraday's most important duties was to make neat copies of Davy's scribbles.

But Davy astonished onlookers in the laboratory by turning bits of apparatus to new uses, thinking with his fingers, and leaping ahead of the spectators in a crucial experiment, proving some unexpected point, or discovering a new substance or property. When Faraday took up his new position, Davy was disabled following an explosion when he was studying compounds of nitrogen and chlorine, and right away Faraday was introduced to the dangers of the laboratory. There were few safety precautions in Davy's laboratory, but on this occasion they wore glass masks, which were indeed damaged in further explosions but saved their eyes from injuries.

In 1813 Davy determined to go to Paris with his new wife, and to take Faraday along; unfortunately at the last minute his valet panicked at the thought of being among enemy aliens and refused to go. Lady Davy chose to regard Faraday as a mere servant, and things were not always easy. But as an introduction to the leading men of science in the world – for Paris was the centre of excellence in most sciences – the tour was a magnificent opportunity for the young man; and to work with Davy, who even on his honeymoon had taken chemical apparatus along, was a superb training. In Paris they were given a sample of a new substance isolated from seaweed; in a race with Gay-Lussac, Davy recognised its analogies with chlorine, and named it iodine. The party went on to Italy; and returned to England, with unique experience of Napoleon's Europe, in 1815. Faraday's horizons had been made much wider; and his abilities were being recognised by Davy, with whom he continued to work in a kind of scientific apprenticeship, by now more like a research student than laboratory assistant.

Back in London, Davy was asked to do something about explosions in coal mines; an increasing problem as workings were made deeper with burgeoning demand for fuel. With Faraday as his assistant, he invented the miner's lamp in which the flame was shielded by wire gauze to dissipate the heat: the investigation showed Davy on top form, using scientific knowledge and experimental intuition in solving a practical problem where rule of thumb could not lead to a reliable device. Davy went for further Continental trips in the following years, leaving Faraday to hold the fort at the Royal Institution and gradually his letters became warmer, so that by 1820 he was writing to Faraday in very much the same terms that he used to his own younger brother John. In 1820 Sir Joseph Banks died. He had been elected President of the Royal Society in 1778 a few days before Davy was born. Davy rushed to England determined to succeed him; and was duly elected. In a classic Dick Whittington story, the country boy made good and occupied the place of the landowning friend of King George III; but such social mobility has a high price, and while Davy was admired and even adulated he had few close friends.

For Faraday too these were the crucial years in which he claimed his scientific territory and achieved his independence. In 1820 H. C. Oersted performed his classic experiment in which a magnetic needle moved when an electric current was switched on and off near it. It had long been felt that there must be some connection between electricity and magnetism, but the exact relationship proved elusive: now all over Europe electromagnetism became a prime topic of discussion and investigation, and Faraday joined in. In 1821 Faraday married Sarah Barnard, a Sandemanian, and became himself a full member of the church; and he was also promoted to Superintendent of the House and Laboratory of the Royal Institution. He was now ready for independence in life and in science; but like other dominating fathers, Davy (his father in science) did not appreciate this, and when Faraday was put up for the Royal Society by somebody else, and was also accused of intellectual trespassing on Wollaston's territory, they had a great row. Like other family quarrels, it was never really made up; which was regrettable for both their sakes.

Faraday afterwards remarked cattily that Davy's example taught him what to avoid. While in background they were not very different, in character they were poles apart. Davy was a religious man, but of a rather vague and liberal kind; he was a great diner-out, sometimes rushing from the laboratory hastily putting a clean shirt over his dirty one, and he enjoyed mingling with the mighty. His ambitions, expressed in a youthful diary when he compared himself with Newton were not merely to understand nature but also to rise in the world; there is nothing immoral in wanting to be President of the Royal Society, but Davy's social superiors (in a very class-conscious world) resented his aspirations and were jealous of his glittering reputation. He thought he had not received sufficient honours; his early friends feared he had been spoiled by success. Faraday, by contrast, had a strong Nonconformist conscience; and his social life was largely among fellow Sandemanians. Seeking God's laws in nature, he was uninterested in power; and while his marriage, like Davy's, was childless, it was, unlike Davy's, very happy. He refused to stand later for the Presidency of the Royal Society, and rejected a knighthood; while very prickly about honour and scientific priority, he did not aspire to the social status of the Davys, with its costs in unease and loneliness. On the other hand, whereas Davy had brought on first his brother and then the young Faraday, in his maturity Faraday took on no young man: his assistant was Sergeant Anderson, who simply did exactly as he was told.

In science, Faraday did not and could not escape from the mantle of Davy. He became an excellent and popular lecturer, founding the Christmas Lectures for children at the Royal Institution and carrying on with the programme of Discourses in the tradition of Davy. He never liked to be called by the new term 'scientist' which seemed to imply narrow specialism; he felt himself to be a natural philosopher. He proved to be an experimental genius; and behind his experiments lay the conviction of Davy that the various forces of nature must be manifestations of one power – qualitative grasping at what we call the conservation of energy. While he succeeded in the 1830s in quantifying the relationship between chemical affinity and electricity to some extent in his famous laws of electrolysis (which describe what happens when an electric current decomposes a solution), he had like Davy a suspicion of mathematics. Armchair science in which one sought to anticipate nature by playing with symbols and numbers was anathema to the experimentalist. Both Davy and Faraday considered themselves to be chemists, and had an uneasy relationship with the Cambridge mathematicians who were coming to dominate the new science of physics in Britain, and whose model for science was astronomy.

In the 1820s Faraday's chief work was in the mainstream of chemistry. He isolated benzene in whale oil in a classic piece of fractional distillation, using a glass tube which he had bent into a zig-zag; and in 1827 published his only book, *Chemical Manipulation*. This was a manual of processes, telling the reader how to weigh, to distil, to grind up or 'triturate', and so on; it is very valuable for the historian who needs to know just what went on in the best chemical laboratories of the day, and can still be read with advantage by the student of chemistry who takes ready-cut filter papers and quickfit glassware for granted. In the days before Pyrex glass, apparatus had to be thick to be strong, and thin to stand heat, and Faraday has advice for coping with this problem; heating was done by furnaces or by spirit lamps. The chemist had to be very good with his hands. Faraday was insistent that nothing should be thrown away; by making do and mending, and by using old pieces of apparatus in new ways, one could get along without waste.

At this period, Faraday was establishing a wide 'professional' reputation as a chemical consultant and analyst, for substantial fees which rose to £1000 a year; rather than employ scientists, British industrialists tended to use consultants, and established men could do very well. The Royal Institution, and other educational bodies, could, by contrast, pay only small salaries to men of science; consultancy was necessary if one wanted to keep up with doctors and lawyers. Faraday also, at the request of the Royal Society, embarked upon two pieces of technological research, on optical glass and on steel. Both these were inconclusive, though they did produce some glasses of very high refractive index and some interesting alloys. Short-term utilitarian research, and routine analyses, did not lead to anything very interesting and left Faraday unsatisfied.

In the 1830s, after Davy's death, he set out on a new course of fundamental research which was ultimately to prove of enormous utility: he began on the series of what he called Experimental Researches in Electricity. Each paragraph was numbered in this great work, which came out in a series of papers for the Royal Society's *Philosophical Transactions*, later reprinted in three volumes; on the frontier of chemistry and physics, Faraday had found his place. Whereas Newton's great work had been *Mathematical Principles of Natural Philosophy*, Faraday's work was experimental; contemporaries found themselves astonished at the discoveries Faraday made, while their mathematical models were as sterile as the epicycles of the old astronomy. They came to believe that he had worked without metaphysics, and almost without theory, which was not the case.

Where most contemporaries saw atoms of matter with void space between them, acting on each other at a distance like the Earth and the Moon, Faraday saw space as filled with lines of force. He came to use the word 'field' for the arrangement of lines of force about a magnet or an electrically charged body; and for him the concept of massy atoms, whether in chemistry or in electromagnetism, came to seem irrelevant and misleading. It was the forces in the space between them which was interesting; and by using iron filings one could for example illustrate the force field around a magnet. Faraday thus hoped to escape the paradox, in which men of science since Newton had found themselves, of explaining the familiar facts of gravitation, chemical affinity, and magnetism in terms of the unfamiliar and incomprehensible notion of matter acting at a distance across void space. For Faraday, matter did not have sharp boundaries; it filled a field. As the great physicist James Clerk Maxwell later put it, matter occupied space like an occupying army with rifles, and not like a mass of men standing shoulder to shoulder.

Faraday's field concept was gradually refined in his experimental interchange with nature; for he was one of the greatest experimenters, and used experiment not just to test preconceived theories but as a form of scientific thinking. He recorded his work systematically in his scientific diary, which has now been published, and which shows his thought maturing. In 1832 he demonstrated electromagnetic induction, interpreting it in terms of his lines of force. He wound coils of wire round the opposite sides of an iron ring; and found that when an electric current is switched on and off in one of them, setting up a magnetic field in the ring, a current is generated in the other. This was in effect a transformer, and Faraday went on to invent the dynamo, in which moving a coil in a magnetic field sets up a current in it. He found conversely that a coil carrying a current in a magnetic field will move, which is the principle of the electric motor. Whenever lines of force were set up and collapsed, or were cut, an electric current or a magnetic field would result. He then demonstrated, as Davy had been trying to do on his death-bed, that the electricity from an electric fish or a charged glass plate or a wet battery are indeed all the same. For Faraday, electricity was not a thing, a sort of juice flowing round a circuit, but a kind of force.

After this intense, original and lonely work, and perhaps from getting too much mercury into his system in the laboratory, Faraday had a breakdown in health in 1839 and did not get fully back to research until 1843: thereafter he complained of loss of memory, but this was not serious until his old age in the 1860s. On 13 September 1845 he recorded in the Diary his success in relating magnetism to light. When light is passed through certain crystals, it is polarised, which is interpreted as having all its vibrations in one plane at right angles to its direction, instead of many as in ordinary light. When polarised light falls onto another similar crystal, it will pass through if the crystal is correctly aligned with the plane of polarisation, but otherwise will be stopped. Faraday passed a polarised ray through a piece of his experimental optical glass: 'It gave no effects when the *same magnetic poles* or the *contrary* poles were on opposite sides (as respects the course of polarized ray) – nor when the same poles were on the same side, either with the constant or intermitting current – *BUT*, when contrary magnetic poles were on the same side, there was an effect produced on the polarized ray, and thus magnetic force and light were proved to have relation to each other. This fact will most likely prove exceedingly fertile and of great value in the investigation of both conditions of natural force.' The plane of polarisation was rotated, so the light no longer went through the second crystal. Faraday was delighted with this success, achieved experimentally after many trials; and he spent a good deal of time in his later years trying to show the analogies of gravitation and electromagnetism, and getting them to act on each other. Here he met with no success.

By this time Faraday was greatly respected throughout the scientific world, and had a wide correspondence. His work had been used by his friend Charles Wheatstone in the first electric telegraphs, laid along railway lines from 1838, and suddenly making it possible to send messages faster than a man on horseback could go. His experimental researches were beginning to be useful; but physicists trained, as most were by the 1840s, in mathematics found his theory hard to grasp or indeed to take seriously. First William Thomson, later Lord Kelvin, the great Glasgow physicist and pioneer of oceanic cables, began the process of giving Faraday's work a mathematical form; but it was Clerk Maxwell who at the end of Faraday's life made field theory and the electromagnetic theory of light central to physics, in his great works of 1864 and 1873. At much the same time the engineering problems of power generation and the conversion of electricity to light in a convenient bulb were being overcome by such men as Thomas Edison and Joseph Swan, the first power stations coming in 1881. So while Faraday was duly appreciated in his lifetime, and widely mourned at his death in 1867, his enormous importance was not fully clear for another generation; there is generally a gap like this between fundamental research and its application. Faraday himself was aware of this, in replying to Sir Robert Peel who had asked what use the dynamo was; he said he did not know, but 'I wager that one day your government will tax it'.

Faraday had given up most of his consultancy work in favour of his electrical research, and was not a rich man when he retired in 1862 to the house at Hampton Court which Queen Victoria made available for him and where he spent his last five years. But he advised Trinity House about lighthouses, was on the Senate of the University of London, and examined at the Royal Military Academy at Woolwich. He worked on colloids, and also on the passage of electricity in gases at low pressures. Here the pumps available were the crucial thing, and what Faraday began was carried on by his admirer William Crookes and then by J. J. Thomson, who in 1897, at Cambridge, 'discovered' the electron

and first publicly described his experiments at the Royal Institution. The very private man, an elder in a small fundamentalist
church, a great lecturer, was not simply one of the great Victorians, but was also crucial in ensuring that we think and live in a
different world from his. Beginning with insights derived from his
long apprenticeship with Davy, in prolonged experimental wrest-
lings with nature he gave us a new understanding of electricity,
magnetism and light, and ultimately the power to use them much
more fully. The irony about his career is that his most useful
research was done when he gave up the hope of immediate utility
in order to find the laws God had given to matter and force.

CHARLES DARWIN:
Solving the Problem of Organic Diversity
John Durant

As the natural sciences progressed, one problem which loomed ever larger was the position of Man within Nature. The 'New Science' argued that Nature was a machine, operating through universal, mechanical laws; Christianity said Man was created with a free will and an immortal, immaterial soul. Science increasingly seemed to rule out miracles, and progress in geology and palaeontology appeared to many to confute the story of Creation as written in Genesis. The works of Charles Darwin (1809–1882) constitute a profound reflection upon the role of Man in the system of things, a role both elevated (only Man understands the operations of Nature), yet also humbling (not only is Man descended from the apes, but in no unambiguous sense can he be seen as Lord of Creation). Many Churchmen and moralists were deeply anxious about the implications of Darwinism for morality, but Darwin thought that by rendering Man part of Nature he had taught the world a truly moral lesson.

Appearances, it is often said, can be deceptive. Certainly this seems to have been so in the case of the young Charles Darwin. In his *Autobiography*, Darwin recalled that as a schoolboy he had not made much of an impression either on his family or on his teachers. The verdict on him in 1825, for example, when at the age of 16 he was taken out of Dr Samuel Butler's School in Shrewsbury and sent to Edinburgh to study medicine, was that he was 'a very ordinary boy, rather below the common standard in intellect'. Indeed at around this time his father, who was a successful Shrewsbury physician, chided him with the memorable words, 'You care for nothing but shooting, dogs and rat-catching, and you will be a disgrace to yourself and all your family.'

This last remark must surely rank very high in the lists of the world's worst prophecies. For the supposed ne'er-do-well Charles went on to become far more successful than his father Robert, and far more famous even than his grandfather Erasmus, the widely celebrated physician, natural philosopher, poet, and inventor. Acknowledged in his lifetime as the brilliant discoverer of some of the most fundamental laws of living nature, and honoured at his death by burial in Westminster Abbey, Charles Darwin is now generally regarded as the most important and influential biologist of his own, indeed of any age.

Darwin was a Victorian gentleman-naturalist. Born into a wealthy and distinguished English family, he developed in childhood both the love of nature and the passion for collecting – 'all sorts of things, shells, seals, franks, coins, and minerals' – that were so characteristic of his century, his country, and his class. Having inherited a substantial fortune from his father, he had no need to go into one of the recognised professions; and this, coupled with a stable and extremely happy marriage to his first cousin Emma Wedgwood (with whom in due course he had ten children) gave him the freedom and the security to pursue the one great intellectual passion of his life, which was philosophical natural history.

Philosophical natural history was the systematic study of plants and animals, both living and extinct, with a view to learning more about their natures, their origins, and their inter-relationships. This was a field of inquiry which British scientists dominated for much of the nineteenth century. In part, at least, their dominance was a consequence of the fact that success as a philosphical naturalist depended far more upon a sharp eye, a vivid imagination and a logical mind than it did upon any very elaborate research facilities such as complex apparatus and large laboratories; and whereas British scientists in the nineteenth century possessed their fair share of the former qualities, generally speaking they lacked the sort of institutional support upon which the newer, experimental sciences were coming to depend.

In the early nineteenth century philosophical natural history was a dynamic and exciting field of inquiry. Between 1800 and 1830, geologists learnt how to interpret the apparently confused jumble of rocks covering much of the earth's surface as an orderly record of events, that had occurred through unimaginably vast periods of time; and as they pieced together the story of the past, one startling fact soon emerged: the history of the planet was marked by apparently endless and often dramatic changes. Down the ages, continents and oceans had appeared and disappeared; climates had shifted from warm to cold and back again; and an almost bewildering array of plant and animal forms had appeared and (generally) disappeared, never more to return. Understandably, the interest of most philosophical naturalists centred on the question of how and why these changes had occurred.

Naturally, some events were easier to account for than others. For example, most philosophical naturalists were agreed that extinction was the result of unfavourable environmental change. Faced with deteriorating conditions of existence, it was widely supposed that particular populations had simply dwindled in numbers until they died out altogether. But of course extinction was only one side of a geological coin the other side of which was the appearance of new species; and here, no such straightforward explanation was available. Quite simply, the question of where new species came from was the biggest problem in all of philosophical natural history. It was what Darwin himself referred to as 'the great fact – that mystery of mysteries – the first appearance of new beings on this earth'.

The philosophical naturalists of the 1830s adopted a variety of approaches towards this 'mystery of mysteries'. Some saw new species as the direct and immediate handiwork of God; some saw them as the result of unknown providential natural laws; and some steered safely around the problem altogether by making vague references to the 'creation' of new species without ever specifying exactly what they meant. Significantly, the vast majority of philosophical naturalists conceived of the problem of origins in frankly religious terms. In particular, they regarded the exquisite adaptations of plants and animals as clear evidence of divine design. According to the widely read natural theologian William Paley, for example, just as the very existence of a watch proved the prior existence of a watchmaker, so the very existence of complex structures such as eyes and wings proved the prior existence of God.

This was the view of life that Charles Darwin absorbed as a young man. Following his indifferent performance at school and his evident dislike of medicine at Edinburgh, Darwin had gone up to Christ's College, Cambridge, in 1828 in order to take a BA degree in preparation for the Anglican ministry. Once again, however, he thoroughly disliked most of his official studies and devoted the greater part of his energies to natural history. In Edinburgh, he had become friendly with the zoologist Robert Grant. In Cambridge, he soon sought out men such as John Stevens Henslow, the Professor of Botany, and Adam Sedgwick, the Professor of Geology; and in their company he underwent an informal but effective training as a biologist and a geologist.

By now, Darwin's love of nature and his passion for collecting (he was especially keen on beetles) had grown into something more like ambition; and under the inspiring influence of the great explorer-naturalist Alexander von Humboldt he developed what he later described as 'a burning zeal to add even the most humble contribution to the noble edifice of Natural Science'. Towards the end of his time at Cambridge he made some preliminary inquiries about the possibility of going to Tenerife; but these were soon abandoned when Henslow, hearing that the Captain of the Admiralty survey ship HMS *Beagle*, Robert Fitzroy, was looking for a gentlemanly companion to accompany him on his forthcoming voyage of exploration, recommended Darwin for the job.

The voyage of the *Beagle* transformed Darwin from a promising novice into a master practitioner of philosophical natural history. Moreover, it set him firmly on the path towards his most important discoveries. Crucial to Darwin's success was the fact that the first volume of the eminent Scottish geologist Charles Lyell's great work *The Principles of Geology* (1830) appeared in time for him to take it aboard the *Beagle*. Lyell had decreed that in matters geological the present is the key to the past, and throughout his five-year voyage Darwin found repeated opportunities to put this precept into practice. In 1835, for example, he witnessed the terrible effects of a major earthquake in Chile. This devastating upheaval permanently raised the level of some of the land around the city of Concepcion by as much as ten feet, and Darwin used this fact to explain how vast numbers of fossil sea-shells at Valparaiso could have ended up at elevations of over a thousand feet.

A few months later, Darwin found himself in the Galapagos Archipelago, a small group of geologically relatively young volcanic islands situated on the equator a few hundred miles off the west coast of South America. Here he was struck by the fact that animals as varied as mocking-thrushes and lizards, ground-finches and giant tortoises were all entirely new species, found nowhere else outside the archipelago. Once again, Darwin reasoned from the present to the past, but this time his deductions led him to a most un-Lyellian conclusion. Clearly, the Galapagos Archipelago had originally been thrown up out of the sea by volcanic eruption. Equally clearly, it had then been colonised principally from mainland South America. However, the present-day inhabitants of the Galapagos were specifically distinct from those of the mainland; and it appeared to follow therefore, that the two descendant groups must have diverged from common ancestors since the time when the islands were first colonised.

Darwin did not leap straight to this momentous conclusion. But as he reflected on his findings during the latter stages of his voyages, and particularly as he assessed the significance of his Galapagos collections following his return to England in 1836, he became more and more convinced that new species originate from old ones by a process of descent with modification. The key question was: how did this process occur? In 1837, Darwin began systematically to assemble ideas and evidence on this subject in a series of private notebooks. Very rapidly, his project expanded to embrace the origins not only of plants and animals but also humankind. As he put it to himself in January 1838, 'from our origin in one common ancestor we may be all netted together'.

Early on, Darwin recognised that the familiar practice of plant and animal breeding might offer clues to the origin of species. He knew that domesticated plants and animals had been bred from wild ancestors in a process involving descent with modification. In this case, the process worked because plant and animal breeders repeatedly selected individuals that just happened to possess this or that desired characteristic. Anxious to learn more about this process, Darwin made himself something of an expert on pigeon-breeding; and he soon came to realise that, given a little time and patience, artificial selection could produce astonishing results. Was it possible, perhaps, that there existed a similar, but entirely automatic, process of selection in nature?

In the autumn of 1838 Darwin found the answer to this question in his reading of the Reverend Thomas Malthus's famous *Essay on the Principle of Population* (first edition, 1798). Malthus had argued that there was a universal tendency for human populations greatly to exceed their available food supplies, with the result that human life was a continual 'struggle for existence'. This idea was a commonplace in the early-nineteenth century, but when applied to the living world in general it was usually assumed to act conservatively, keeping individuals or entire species 'up to the mark' of their ideal form. By the time that he encountered Malthus, however, Darwin was well on the way to abandoning the notion of ideal form; and thus he was able to see in the struggle for existence not a conservative principle but rather a radical force for change. If throughout nature more individuals were continually being born than could survive to reproduce, and if among these individuals there existed variations that could be passed on by inheritance, then any individuals which happened to possess useful variations would tend to be favoured in the struggle for existence; and in this way such variations would accumulate down the generations in much the same way as did desirable variations in domesticated races of plants and animals. With the analogy with artificial selection very much in mind, Darwin called this process *natural selection*.

From the outset Darwin saw natural selection as the key to the great problem of the origin of species. At a stroke, it explained not only how species changed but how they changed *adaptively*, that is, in ways fitting them better for survival and reproduction in their natural environments. But having hit upon natural selection in 1838, Darwin waited almost four years before permitting himself the luxury even of a private pencil-sketch of his theory, another two before producing a 'brief abstract' in ink, and twelve more before finally setting to work to write up his ideas for publication. Throughout this period Darwin kept his theory a closely guarded secret among the members of his immediate family and a small circle of close and trusted friends, including Charles Lyell and the botanist Joseph Hooker.

Why all this secrecy and delay? The answer, of course, is that Darwin knew just how controversial his views were likely to be. From the outset, he had been part of the tradition which regarded the problem of the origin of species as being bound up with religion. At Cambridge, he had read William Paley's work as part of the BA curriculum; and he had found Paley's argument from design for the existence of God utterly compelling. Now, however, natural selection undermined that argument; and Darwin the Victorian gentleman-naturalist knew that it would be widely seen as an attack, not just upon one particular form of natural theology, but upon the very foundations of the Christian faith.

Darwin's position was extremely difficult. At Cambridge, he had been more-or-less orthodox in his religious beliefs. In the 1830s, however, disbelief, in his words, 'crept over me at a very slow rate, but was at last complete'. The reasons for this disbelief were complex – part historical, part moral, part scientific – and its effects were traumatic. From a personal point of view, there was the question of Darwin's relationship with his wife Emma, who was a devout Christian; and from a professional point of view, there was the problem of causing offence amongst his colleagues, many of whom attached great importance to the apparently secure supports offered to orthodox religion by philosophical natural history. Not surprisingly, perhaps, Darwin's private notebooks for 1838 refer to the 'persecution of early Astronomers', as well as to dreams of execution. Later, he was to tell his close friend, the botanist Joseph Hooker, that to admit his theory of the origin of species was like 'confessing a murder'.

The final ingredient in Darwin's great reticence to publish was his intense dislike of any kind of public dispute. Increasingly after 1840, Darwin's general health was rather fragile (theories abound, but we shall never know for certain whether he was suffering the effects of a tropical disease caught during the *Beagle* voyage or whether other, possibly psychological, causes were at work); and for the remainder of his life he depended utterly upon the tranquility of domestic life at his home at Down in Kent. Darwin, the great theoriser of the struggle for existence, was not keen to engage in a personal struggle for the existence of his theory.

Thus it was that Darwin found himself still working away privately on the manuscript of a big book on natural selection in June 1858, when he received a letter from the naturalist Alfred Russel Wallace containing a brief but perfectly accurate statement of his almost twenty-year-old views. Wallace was a younger and less well-established naturalist who had journeyed to South America in the late 1840s and was now travelling around the islands of Malaysia in search of clues to the origin of species. Intellectually as well as geographically, Wallace's work had paralleled Darwin's to an extraordinary degree; and now he too had hit upon the idea of natural selection and realised its great significance. Wallace had been corresponding with Darwin on other matters; and so it had seemed sensible to send Darwin a short account of his idea to see what he made of it!

Darwin was devastated. 'So all my originality, whatever it may amount to', he wrote to Lyell, 'will be smashed, though my book, if it will ever have any value, will not be deteriorated; as all the labour consists in the application of the theory.' In the event, it was Lyell and Hooker who secured Darwin's originality by arranging that a selection of his letters and manuscripts describing natural selection should be read together with Wallace's paper as a joint contribution before the Linnean Society of London. In the end, also, Darwin's prediction concerning the long-term value of his work was borne out; for abandoning his big book, already several years in preparation, he settled down to write a much shorter work for immediate publication. Finished in a matter of months, and published before the end of 1859, this work established Darwin's title as the first and most brilliant exponent of the theory of evolution by natural selection.

On the Origin of Species by Means of Natural Selection, or the Preservation of Favoured Races in the Struggle for Life was a remarkable *tour de force*. Presented to the public as an 'abstract', with no references of any kind, it was an easily readable, powerful, and sustained argument in favour of a radically new view of life.

Throughout, Darwin had to contend with two great problems. First, by its very nature his theory was not open to definitive proof. The evidence in favour of evolution by natural selection, though strong, was largely circumstantial; and in places it faced genuine technical difficulties. Second, the theory flew in the face of some of his colleagues' most deeply held beliefs: about the distinctness and the inviolability of species; about the significance of organic adaptation; about the purposefulness of the history of life; and, last but not least, about the physical and moral status of humankind.

Darwin exercised great skill in overcoming these problems. While not pretending that he could prove his case conclusively, he was at pains to demonstrate two things: first, that natural selection could have been the principal cause of organic diversity; and second, that on this supposition, and on this supposition alone, all manner of hitherto puzzling and unrelated facts about the living world were readily explained. A master-stroke of argumentation was Darwin's frankness in uncovering all kinds of apparent difficulties for his theory, only to show why these were not genuine problems at all. After 1859, very few critics were able to find faults with the theory of evolution by natural selection that Darwin himself had not already foreseen and forestalled in the *Origin*.

When it came to dealing with the great mass of prior beliefs which stood in the way of his theory, Darwin was equally skilful. For one thing, he intentionally excluded all but the most fleeting references to what he knew was the most controversial aspect of his theory, namely its implications for human origins; and for another, he took great pains to point out that his theory was not in principle anti-religious. After all, natural selection was simply an extension of the domain of natural law to cover the origin of species; and had not philosophical naturalists long been accustomed to regarding natural laws as a mode of divine providence? 'To my mind', Darwin wrote in his final chapter, 'it accords better with what we know of the laws impressed on matter by the Creator, that the production and extinction of the past and present inhabitants of the world should have been due to secondary causes, like those determining the birth and death of the individual.'

Skilful as this and similar devices undoubtedly were, they could not avert the storm of controversy which greeted the publication of the *Origin* on 24 November 1859. The day before publication, Darwin's young supporter Thomas Henry Huxley wrote to warn him of 'the considerable abuse and misrepresentation which, unless I greatly mistake, is in store for you', and promising his active assistance: 'I am sharpening up my beak and claws in readiness.' In the event, Huxley's prophecy proved accurate, and his formidable 'beak' and 'claws' were soon busy on Darwin's behalf. His most famous encounter took place in the summer of 1860, when he clashed in open debate with the anti-Darwinian Bishop of Oxford, Samuel Wilberforce, at the annual meeting of the British Association for the Advancement of Science. Faced with an apparently insulting inquiry concerning his own descent from an ape, Huxley retorted that he would rather be descended from a 'miserable ape' than from 'a man highly endowed by nature and possessed of great means of influence and yet who employs those faculties and that influence for the mere purpose of introducing ridicule into a grave scientific discussion.'

Huxley's riposte had just the right combination of righteousness and outrageousness to capture the Victorian imagination. He was, he wrote to a friend, 'the most popular man in Oxford for full four and twenty hours afterwards'. Darwin looked on from the safety of his study with faintly anxious amusement: 'How durst you attack a live Bishop in that fashion? I am quite ashamed of you! Have you no respect for fine lawn sleeves? By jove, you seem to have done it well!'; and before long, the encounter had grown into a legend. According to one story, when the Bishop of Worcester told his wife of the affair, she is supposed to have said, 'Descended from the apes! My dear, let us hope that it is not true, but if it is, let us pray that it will not become generally known.' After Oxford, it took some time for calmer and quieter voices to point out what was implied in the *Origin* itself, namely that Christianity was no more to be equated with Paley's view of the origin of species than it was to be identified with Ptolemy's view of the sun and the planets.

The impact of the *Origin* both on science and on wider society was enormous. Scientifically speaking, the book was decisive in converting a majority of biologists to the idea of evolution, even though it did not immediately persuade all of them that natural selection was its principal mechanism. After 1860, old-fashioned philosophical natural history went into sharp decline. Increasingly, biologists abandoned the theological principles of design and purpose in favour of the naturalistic principles of inheritance and variation. Of course, the revolution was not complete overnight. In the late-nineteenth century many even among Darwin's supporters continued to incorporate theological elements within evolutionary theory, especially in connection with the problem of human origins (Alfred Russel Wallace himself is a case in point);

but after 1859 the centre of gravity of biology shifted, and it shifted for good.

The social impact of the *Origin* was equally great. For a time, some naturalists agonised over the religious and the moral implications of descent with modification; but very rapidly such doubts gave way to what can only be described as a wave of Darwinian enthusiasm. Sweeping aside the fears and protests of an earlier generation, the naturalists of the 1860s almost fell over themselves in their zeal to apply the theory of evolution by natural selection to human affairs. By the time that Darwin published *The Descent of Man* in 1871, Huxley, Lyell and Wallace in England, and Ernst Haeckel and Karl Vogt in Germany had already beaten him to it. Now, the question was not whether evolution was true, but how it could be made to yield momentous conclusions about human nature, human origins, and human destiny.

Prominent among the conclusions that were drawn from Darwin's theory in the late-nineteenth century was the idea that the struggle for existence – between individuals, social groups, institutions, nations or even races – was a necessary and/or a beneficial law of human history. For a time, 'social Darwinism' became the most fashionable of social gospels. Of course, Darwin was not the only influence at work here. Many other writers in the 1850s and 1860s were developing a broadly evolutionary perspective, not just in biology, but in anthropology, history, psychology, and social theory as well. The English philosopher Herbert Spencer, for example, single-mindedly applied the idea of evolution to absolutely everything. He it was who coined the term 'survival of the fittest', and much that passed for 'social Darwinism' might better have been termed 'social Spencerism'.

There was a little, but only a little, justification for 'social Darwinism' in the writings of Darwin himself. For although Darwin was adamant that he did not believe in any universal law of progress in nature, yet he did speak in the *Origin* of natural selection causing 'all corporeal and mental endowments' to 'progress towards perfection'; and although he wrote to Lyell that he had read in a Manchester newspaper 'rather a good squib' that he had proved 'might is right, and therefore that Napoleon is right, and every cheating tradesman is also right', yet he went on to speculate in the *Descent* about the evil effects of the cessation of the struggle for existence in civilised societies; and in 1881 he remarked to one correspondent that 'I could show fight on natural selection having done and doing more for the progress of civilisation than you seem inclined to admit . . . The more civilised so-called Caucasian

races have beaten the Turkish hollow in the struggle for existence.' Darwin was a Victorian gentleman-naturalist, and he was
by no means free of all the presumptions and prejudices which
went with the position.

Nevertheless, there was something profoundly out of keeping
with the spirit of Darwin's work in the almost reckless use which
was made of his ideas by ideologues in the late nineteenth century.
For above everything else Darwin was an endlessly curious naturalist who delighted in the pursuit of a better understanding of the
living world. After 1859, and while many others were carried away
with the most fanciful and far-fetched evolutionary speculations,
Darwin continued patiently with his scientific studies in a dozen
different fields. For example, he applied his theory of natural
selection to the pollination mechanisms of flowers; he carried
forward his studies of inheritance and variation; and he worked
productively on the problems of human origins and emotional
expression. Appropriately enough his last book was a detailed
study of the habits of earthworms.

Of course, not everything that Darwin touched turned to gold.
For example, despite great efforts he failed satisfactorily to solve
the great problem of inheritance. Darwin developed a theory of
'pangenesis', according to which each part of the body produces
minute 'gemmules' that pass through the bloodstream to the sex
organs. He suggested that gemmules were formed during life, that
they incorporated both inherited and acquired characteristics,
and that by a process of blending they reproduced the appropriate
body parts in offspring. In fact, none of these ideas ultimately
proved useful. A far better solution to the problem of inheritance
was developed by Darwin's contemporary, the Austrian monk
Gregor Mendel; but the value of Mendel's theory, according to
which discrete factors ('genes') recombine during sexual reproduction without blending and without contamination from
acquired characteristics, was not recognised until around 1900;
and even then it took several more decades before Darwinism and
Mendelism were at last brought together in the so-called 'Synthetic Theory of Evolution'.

Despite this signal failure, Darwin's track record as a theoretical biologist was astonishingly good. Time and again, his ideas
went to the heart of the matter. A good case in point was his theory
of sexual selection. This was an extension of natural selection to
explain characteristics that gave animals specific advantages over
their competitors in the struggle for mates. Such characteristics,
Darwin suggested, were of two basic types: weapons (such as the

stag's antlers), which were the products of 'male battle'; and ornaments (such as the peacock's tail), which were the products of 'female choice'. In the late nineteenth century sexual selection by female choice was widely ridiculed; and despite an important contribution by the mathematical geneticist Sir Ronald Fisher in the early decades of this century, it was only around 1970, fully a century after Darwin's original exposition, that substantial numbers of evolutionary biologists finally came to see it as a potentially important evolutionary mechanism.

Today Darwinism is the key to our understanding of the mystery of life. Of course, some people still share the view of many early Victorians that there is something dangerous or even degrading in the very idea of evolution by natural selection; but for those whose sense of identity is secure enough to accommodate the notion that 'we are all netted together', for those whose sense of self-esteem is robust enough to cope with the idea that we are the products of natural selection, and for those whose religion is tolerant enough to admit the truth of things not known to the ancient Israelites, there is the possibility of a very different view. As Darwin himself put it in the final sentence of the *Origin*, 'There is grandeur in this view of life, with its several powers, having been originally breathed into a few forms or into one; and that, whilst this planet has gone cycling on according to the fixed law of gravity, from so simple a beginning endless forms most beautiful and most wonderful have been, and are being evolved.'

LOUIS PASTEUR:
In Pursuit of the Infinitely Small
William Bynum

The age of Darwin saw the life scientist as the gentleman amateur, living in rural seclusion. Within a few years, the meteoric rise of and public acclaim for Louis Pasteur (1822–1895) demonstrated how rapidly in the late nineteenth century science was becoming the focus of attention and national aspirations. Pasteur's microscopic experiments demonstrated the existence of micro-organisms (bacteria) whose spread could be reliably linked to epidemics of cholera, typhoid and other rampant killer diseases. Once the agents of disease had been identified, control, diagnosis and counter-measures became more feasible. Pasteur had a flare for showmanship, but in the long run he achieved fame not because of his publicity-mindedness but because his scientific work was of such fundamental importance.

Few scientists have ever conformed better than Louis Pasteur to the image which their contemporaries held of the ideal scientist. By the time he died in 1895, he was a figure of international veneration, 'the most perfect man who has ever entered the kingdom of science'. To be sure, there were some who had looked upon the growth of science during the nineteenth century with suspicion and fear, to whom science spelled the erosion of traditional religious values and their replacement with a gospel of materialism and the cruelty of animal experimentation. But to most, Pasteur had become living proof of all that was good in science and its application in furthering economic prosperity, promoting health and eradicating disease. That he also seemed modest, pious and devoted to his family, friends and students gave him an aura of saintliness. He died with one of his hands in that of his wife, his other clutching a crucifix.

Pasteur was, in fact, a good deal more complicated than the myths of simplicity and saintliness depicted. Although modest in his personal habits, he was intensely ambitious in his scientific life, aggressive with colleagues and saw himself as a kind of latter-day magus, grappling with the innermost secrets of life itself. Even as a young man he identified with the great scientists of the past, with Galileo, Newton and Lavoisier, and he never lost a sense of the community of scientists. The community was becoming increasingly transnational, as frequent large international congresses attracted much publicity. Although Pasteur participated in many of these, he remained an intense French patriot, particularly after the humiliations of the Franco-Prussian War of 1870–71, subsequently refusing German decorations and honorary degrees, and indignantly returning a degree which the University of Bonn had awarded him. Science was for him one of the highest creations of the human spirit, but it was also to be an important force in regenerating French society. Pasteur never lost the religious faith of his fathers, but nor did he quail at the idea of a priesthood of scientists.

In other ways, too, Pasteur embraced – and helped create – the forms and values of the emerging scientific culture of his day. His life revolved around the laboratory, which he always viewed as the sanctuary of discovery. As circumstances dictated, his laboratory might be set up in a brewery, farmyard or hothouse. He worked in essence alone, although particularly after his health began to fail him, he relied on a succession of assistants and students to help him with the manual part of his endeavours. He inspired in these younger men loyalty and respect, but he rarely took them into his confidence. He used their hands but not their minds, assigning them experimental tasks without discussing with them the larger research project. Above all, he believed in the power of experiment to uncover facts which, properly interpreted, would increase man's understanding and control of Nature. For much of his career, Pasteur directed his experimental genius primarily towards the elucidation, prevention and treatment of disease, and it was symbolic of the increasing importance of the laboratory for medicine that this Frenchman, so closely identified with that most fundamental of modern discoveries, the germ theory of disease, was not even medically qualified. Appropriately, it was not simply human disease which intrigued him, but the 'diseases' of wine and beer, of silkworms and farm animals.

Pasteur's success lay in the fact that he created a unity out of a multiplicity of research problems. Superficially, the diversity is apparent, for his investigations ranged from crystallography to immunology, from spontaneous generation to the practical issues of industrial production. Beyond the seeming eclecticism and

haphazard chances of his career ('Chance favours only the prepared mind', he once insisted) stand a unity and logic which, in
retrospect, seem almost inevitable. At every turn we find micro-
organisms, and Pasteur's fascination with 'the infinitely great
power of the infinitely small'. It was this power which early caught
Pasteur's attention and to the explication of which so much of his
energy was always directed.

Despite Pasteur's passionate commitment to science, his early
life suggested neither its future direction nor its greatness. He was
born on 27 December 1822, in Dole, Jura, although the family
moved a few miles away to Arbois when he was five and it was in
Arbois where he grew up and which he always considered home.
His father, a decorated veteran of the Napoleonic wars, was a
tanner, as had been his father before him. Pasteur was raised in a
close-knit *petit bourgeois* family, whose values of industriousness,
moral earnestness and sobriety he thoroughly absorbed and never
questioned. An only brother died in infancy, so he was always the
darling of the family, encouraged by his parents and doted on by
his sisters. His youthful enthusiasm was not science but painting
and drawing, for which he had a good deal of talent, but he was
sufficiently accomplished academically to secure a place in the
prestigious Ecole Normale in Paris. By then his scientific aspirations were firm, and from 1843 until 1848 he studied and worked
in the science section of the Ecole Normale, passing his doctorate
in 1847. His career there was distinguished without being brilliant,
his training so broad that chemistry and physics came equally
within his ken.

Throughout his early career, Pasteur looked to older scientists
for inspiration and patronage. In those formative years, this group
of father figures included the chemists Dumas and Balard, the
physicist Biot and the mineralogist Delafosse. From Dumas Pasteur imbibed a sense of dignity of scientific research; in Balard he
found a loyal patron and friend; and from Biot and Delafosse he
received his first major experimental puzzle. It concerned a
phenomenon known as optical activity, the capacity of certain
substances, in crystalline form or dissolved in solution, to cause a
beam of polarised light to deviate to the left or right. One such
optically active substance was tartaric acid, a compound prepared
from some of the waste products of wine-making and important at
the time both in medicine and dyeing. Tartaric acid was of
additional interest because it was known that another compound
existed of apparently identical chemical composition and roughly
comparable chemical properties but lacking any optical activity. It

was called racemic acid, and Pasteur set out to examine the relationship between crystalline form and optical activity of these acids and their salts, the tartrates and paratartrates.

One of his early observations was completely expected: that optically active substances have asymmetrical crystalline shapes. Further work convinced him that the asymmetry of the crystals reflected a more fundamental molecular asymmetry which could be produced only by living things. Synthesise compounds in the laboratory and they display boring regularity. Only substances manufactured by living cells could display the asymmetry necessary to yield optical activity. Tartaric acid, after all, was a product of wine-making, which was sometimes seen as a vital, living process. Occasionally, the optically inactive racemic acid appeared, but Pasteur was able to show that it was composed of two kinds of crystals, each optically active if separated, but not if mixed together in equal portions.

A good many of Pasteur's contemporaries believed that insuperable barriers existed between living and non-living things ('vitalism'), and Pasteur's insistence that asymmetry was uniquely associated with life accorded well with the vitalism of his times. Most vitalists used this belief as a means of setting ultimate limits on scientific knowledge and of assuring a permanent role for the Creator (the source of the vital principle) in the order of things. While Pasteur was never active in the more general debates about the relationship between science and religion, keeping his own religious beliefs private, he always maintained the distinction about the specialness of life and of substances produced by living organisms. Curiously, though, he felt that his own discoveries might make it possible to create life in the laboratory. He speculated that optical activity was produced because the earth itself was subjected to asymmetrical forces. Thus light striking the earth might be asymmetrical, he reasoned, as might electrical and magnetic forces. Consequently, it should be possible to produce asymmetrical substances – and maybe even life itself – in the laboratory, by reproducing these conditions of asymmetry. Some of his colleagues discouraged him from such grandiose aspirations, but Pasteur had some powerful magnets specially constructed and attempted to unlock the innermost secrets of life. No wonder his wife could write that his experiments, 'will give us, should they succeed, a Newton or a Galileo'. These Faustian endeavours failed, and Pasteur himself was subsequently slightly embarrassed by his youthful hubris.

It was part of the underlying unity of Pasteur's work that he never lost interest in the problems which had earlier preoccupied him, and throughout the 1850s he continued to investigate optical activity and molecular asymmetry, devising methods, for instance, for the selective production of several active compounds. This he was able to do because he had become fascinated by the micro-organisms whose metabolic peculiarities led to their appearance. The tartrates were commonly derived from the waste products of wine-making, and by the mid-1850s Pasteur had become intrigued by the causes of fermentation.

'Fermentation' is a general term describing the breakdown, under appropriate conditions, of organic substances. During the process of beer or wine making, the relevant sugars are broken down to form, among other things, ethyl alcohol. When milk ferments, the principal product is lactic acid and the milk goes sour. When ethyl alcohol itself undergoes fermentation, the result is vinegar, the operative component of which is acetic acid. Now, these and many other fermentations are familiar everyday occur-rences, so economically and culturally important that it is hardly surprising that the mechanism of fermentation had attracted a good deal of scientific attention. Both brewing and viticulture had developed as essentially empirical crafts, and though it was known that 'brewer's yeast' needed to be added to make beer ferment, it was not quite clear why this should be so, nor was it understood why the grain or the grape sometimes yielded unpalatable beer or wine.

In the 1850s, scientific orthodoxy stated that fermentation was essentially a chemical process, whereby, for instance, the larger sugar molecules were broken down chemically into simpler alco-hols (or alcohol into vinegar) and by-products such as carbon dioxide. The famous German chemist Justus von Liebig had developed a theory whereby the yeast, in the act of dying, released some kind of vibration which split the larger molecules. The great Swedish chemist Jons Jacob Berzelius had suggested that fermen-tation was to be understood in terms of simple catalysis, the 'ferments' (whatever they were) being capable of speeding up (catalysing) the chemical reactions of fermentation. In any case, those who believed that fermentation was a 'vital' process (pro-duced by living organisms) were in the minority, thought to be inadequately appreciative of the explanatory powers of modern chemistry.

One factor which linked Pasteur's studies on fermentation with his earlier preoccupations was the knowledge that some alcohols are optically active and by his criteria they must be produced by

153

'vital' mechanisms. He was thus predisposed to doubt the authority of two of the most eminent chemists of his time, and by the mid 1850s his work had turned him into something of a microscopist and expert on the life cycles of the microbes involved in tartrate manufacture. Consequently, there are striking continuities between that work and his first major paper on fermentation (1857). In it, he dealt with lactic fermentation, but the contribution contains most of the major features of his mature theory of the role of micro-organisms in the fermentative process. In particular, he analysed the conditions under which he could produce, at will, lactic fermentation from sugar, chalk, brewer's yeast and a little of the grey deposit produced from an earlier lactic fermentation. Under the microscope, this deposit was shown to consist of small 'globules' or very short segmented filaments, and the fermentation to be accompanied by the multiplication of these living microscopic organisms. The lactic 'yeasts' were related to, but not identical with, ordinary brewer's yeast, as evidenced by the slightly different environmental conditions favouring their growth.

Over the next few years Pasteur extended his work on specificity of ferments, establishing firmer criteria for the nutritional conditions necessary to support their growth and activity and demonstrating that alcoholic fermentation was a much more complicated process than the simple conversion of sugar to alcohol and carbonic acid. He also showed that the alcoholic ferment was able to obtain its nitrogen from a non-organic source (ammonium), arguing that what we observe as fermentation was the result of the organism's metabolic activity, as it derived its carbon from the sugar, its nitrogen from the ammonium and its minerals from the incinerated brewer's yeast which had to be added before the fermentation could occur. In late 1860, working with one ferment, he noticed that the rod-like organism lost its motility around the edges of the infusion, remaining active in the centre, where the concentration of oxygen would be lowest. Passing a stream of ordinary air through the liquid actually rendered the micro-organisms immotile, suggesting to Pasteur that not only could this ferment live without free oxygen, but that oxygen in sufficiently high concentration could act as a poison. Unlike yeast, which seemed to be a form of plant life, these 'infusoria' (as he called them) had the motility of animals: the first known example of an animal ferment and of an animal which could live without free oxygen.

This led Pasteur to inquire more closely into the effects of oxygen on other fermentations, and to discover that when free oxygen was plentiful, brewer's yeast thrived in a sugar solution, but alcohol was not produced. Take away the oxygen, and alcoholic fermentation took place. Fermentation, he was to conclude, was 'life without air', as the organisms turned to sugar as a source of oxygen in its relative absence in the free state. The discovery of what Pasteur was eventually to call anaerobic ('without air') organisms also gave him insight into the similarities between fermentation and putrefaction (between, for instance, the souring of milk and the rotting of meat). Both, he insisted, are the result of micro-organisms; both involve the breakdown of larger organic molecules into smaller and also inorganic ones; and both are important dimensions of the economy of nature.

'There are no such things as pure and applied science – there are only science, and the applications of science', Pasteur frequently said, and his work on fermentation amply demonstrated his epigram. Theoretically of great significance, in his hands it quickly led to practical applications. His studies of vinegar manufacturing in France, beginning in 1861, showed that what was traditionally known as 'mother of vinegar' (the lees of vinegar fermentation) contained a micro-organism which Pasteur identified, establishing after much effort the nutritional conditions under which it could be grown and used at will in the production of vinegar. Even more important economically were his studies on wine and its 'diseases'. 'Ageing' of wine, he showed, was accomplished by the slow absorption of oxygen through the porous wood of the casks; its 'diseases', whereby the wine could acquire various unpalatable characteristics, resulted from the presence of unwanted micro-organisms. He experimented unsuccessfully with the addition of various 'antiseptics', to control these unwanted micro-organisms, but he was able to report in 1865 that the 'germs' of wine could be controlled by heating the wine in closed vessels for an hour or so at a temperature of about 60°C. This killed the foreign micro-organisms without damaging the taste, bouquet or colour of the wine. This was a description of the process which within two or three years was being called 'pasteurisation', still such an important procedure for preserving milk and many other products. He had not been the first to recommend heating wine in order to prevent it from spoiling, and his patenting of the process landed him in a rather undignified priority dispute. He was, however, the first to explain *why* heating could preserve wine, and his advocacy of it led to its widespread adoption.

By the 1860s Pasteur was a scientist of international repute, but his ideas on the role of micro-organisms in the genesis of optically active substances, in putrefaction and fermentation, and in the 'diseases' of wine were not universally accepted. Sometimes his experiments could not be duplicated by others, and often his results could be interpreted in alternative ways. Pasteur showed himself adept at the propaganda and rhetoric of science, and not just in its practice. He seemed to enjoy controversy, and friends sometimes lamented the energy he spent attacking opponents. Many of his scientific adversaries were French, but he particularly relished a good fight with Germans like Liebig with whom he debated the nature of fermentation until the latter's death in 1873. But perhaps no debate stirred up so much public excitement as that with his countryman Felix Pouchet, on the issue of spontaneous generation.

Spontaneous or 'equivocal' generation refers to the ancient belief, periodically enunciated in modern times that under certain conditions, living organisms can be formed from non-living materials. The classic example is the appearance of maggots on rotting meat, and since the development of the microscope in the seventeenth century naturalists had seen unicellular organisms which were reasoned to have appeared spontaneously, perhaps from the recombination of putrefying organic matter.

By the early 1860s, Pasteur knew a good deal more about the habits, structures, functions and reproduction of micro-organisms, and it was part of his insistence on the integrity of vital processes that these 'germs' were introduced into the situations where they were active, rather than being produced by such phenomena as putrefaction or fermentation. The dust in the air was, he already had reason to believe, a frequent vehicle of contamination. Consequently, when Pouchet published (in the same year as Darwin's *Origin of Species*) a long treatise on what he called 'heterogenesis', Pasteur was piqued into action.

Pasteur probably never did a more famous series of experiments than those aimed at proving that spontaneous generation does not occur. In particular, the swan-necked flasks which he used have become associated with his image as an experimentalist. To be sure, these investigations demonstrated Pasteur's superb technical abilities, his capacity to design telling experiments and achieve consistent results. He learnt in the process a good deal about the physical conditions of time, temperature and pressure necessary to kill micro-organisms, about the importance of dust in the atmosphere as a mechanism in the spread of germs, and about the relative purity of air at different altitudes. He also had much the better of the debate with Pouchet, convincing most of the scientific community that organisms never appear spontaneously, but result from the reproduction of pre-existent organisms. Life could come only from life. Ironically, however, Pasteur's success in this dispute owed much to his polemical vigour rather than the unequivocal preciseness of the experimental data, for the presence of heat-resistant spores in some of the infusions which Pouchet thought he had sterilised in his flasks, meant that, within the knowledge available to both men, spontaneous generation was a reasonable interpretation of his findings. Pasteur was convinced that Pouchet's results came merely from sloppy techniques.

With the benefit of hindsight, we can see that Pasteur let his beliefs about spontaneous generation influence his willingness to dismiss Pouchet's work as invalid and careless. Pasteur approached this debate with his ideas fixed and none of the equivocal experimental results made him waver from his position. In the other main line of inquiry which he undertook in the 1860s, however, Pasteur showed himself to be far more flexible and open in his attitudes. For more than a decade, the French silk industry had been threatened by a condition called *pébrine* which was killing large numbers of silkworms. The economic consequences were so serious that a governmental commission had been set up under Pasteur's old friend, Dumas, and the latter asked Pasteur to investigate. Pasteur accepted the challenge, and turned himself into an expert on the nuances of silkworm breeding. Much of this work was carried on under trying personal circumstances (he lost his father and two daughters within about a year), but Pasteur, his wife, remaining daughter and several assistants were able to develop, with the aid of the microscope, a series of rough and ready guidelines which could accurately predict whether the larvae would develop into the productive silkworm stage. By destroying the unhealthy larvae, it was possible to control both *pébrine* and another disease of the worm, *flaccherie*.

Despite the fact that an Italian scientist named Bassi had earlier demonstrated that yet another silkworm disease was infectious, Pasteur originally believed that the diseases he was studying were developmental rather than infectious, and he devised his pragmatic preventative techniques without a clear idea that these diseases might be caused by micro-organisms, or spread from worm to worm. Further work, however, enabled him to identify and grow the bacterium which causes *pébrine*, giving him solid evidence for a germ theory of disease. He had long believed that micro-organisms can cause disease and this gave him the impetus to turn to a more systematic investigation of the role of germs in the diseases of man and farm animals.

He was not, of course, the first to come to believe that typhoid fever and malaria, for instance, are distinct and specific diseases, and that this specificity arises from a specific cause. He was not even the first to suggest that these specific causes are living beings such as the bacteria and fungi he had long studied. But his previous scientific work gave him the capacity to contribute creatively to the establishment of the germ theory of disease, beginning with a disease which his countryman Casimir Davaine had shown in the 1860s might be caused by a bacterium. The disease was anthrax, primarily an infectious condition of farm animals (although occasionally contracted by human beings), and convenient to study because the blood of animals dying from anthrax actually contains the rod-shaped anthrax bacillus. In 1876 the young German bacteriologist Robert Koch published his classic work on the life cycle of the bacillus, showing how the infection can spread or can lie dormant for long periods before breaking out again. Pasteur himself quickly extended these studies, showing for instance, that hens, normally not susceptible to anthrax, could be given the disease if their body temperatures were lowered by keeping them in cold water. Conversely, animals so infected could be 'cured' by raising their body temperature. Almost from the beginning, Pasteur was interested in the preventative and therapeutic potentials of his research, suggesting that some infectious diseases might be treated by heating the infected animal or human being.

By the late 1870s Pasteur was energetically extending the germ theory to other diseases, including puerperal fever, osteomyelitis (a bone infection) and boils. He also believed that it should be possible to produce vaccines for these diseases, analogous to Jenner's vaccine for smallpox, which had been known for almost a century. His first success along these lines came with a disease of chickens called chicken cholera (which has no relation to human cholera). By growing successive cultures of the microbe in chicken broth, he was able to produce an organism which was less likely to kill a chicken than usual. Although he usually renewed the broths each day, he returned from a holiday to the laboratory to discover that the microbes had died, the broth was sterile and, when injected into chickens, produced no disease. This had not shown anything except that the cultures no longer had any virulent organisms in them. But with that chance which always seemed to favour Pasteur's prepared mind, he went one step further. He decided to re-use the chickens in additional experiments. He injected them with a quantity of fresh cholera microbes, known to be virulent. The chickens survived. He had prepared an artificial vaccine.

Pasteur demonstrated his findings in 1880 to the Academy of Medicine. By then, however, bacteriology was becoming public news. The City of Paris gave him a site for new laboratories and stables for experimental animals, and by 1881, he had developed a vaccine for anthrax. Always prepared to demonstrate his mastery to a wide audience, Pasteur arranged for a public trial to be conducted at Pouilly-le-Fort. Even *The Times* sent a reporter. Twenty-four sheep, one goat and six cows were inoculated with his attenuated anthrax culture on 5 May 1881, and again on 17 May. On 31 May he and his assistants then injected them, along with twenty-nine unvaccinated animals, with a virulent anthrax culture. On 2 June, when the public was invited to observe the consequences, all of the vaccinated animals were alive, all but three of the unprotected sheep already dead (and two of the remaining ones conveniently expired before the spectators) and the four unvaccinated cows showed signs of anthrax. It was a dramatic triumph, although it left Pasteur with the less exciting tasks of trying to control standards for producing additional vaccine, of worrying about occasional inoculation fatalities, and of squabbling with Koch and others about the mechanisms whereby immunity was produced. Pasteur did have the satisfaction of seeing the Prussian ministry of agriculture advocate his vaccine. A vaccine for swine erysipelas followed within a couple of years.

The last major disease which Pasteur studied was the most complicated of all: rabies. We now know that this most dreaded of diseases is caused by a virus, and that viruses are too small to be visualised by the microscopes available to Pasteur. Nor could the rabies virus be cultivated in the kinds of broths and mixtures he was used to working with. Despite these and other obstacles, Pasteur managed to develop a series of techniques whereby he could attenuate or potentiate the virus (even though he never knew exactly what the causative agent of rabies was). By 1884 he and his colleagues were sufficiently skilful at handling the virus for talk of a vaccine to seem appropriate. Early the following year he was able to treat dogs bitten by rabid animals, and though the step from dogs to human beings was a fearful one, he agreed in July 1885 to treat a nine-year-old boy, Joseph Meister, who had been attacked by a dog feared to be rabid and who, according to his doctors, was doomed to inevitable death. Carefully monitored by members of the French rabies commission, Pasteur inoculated Meister with a series of rabies vaccines of increasing potency, and the young boy never developed the fatal symptoms of hydrophobia. Pasteur began treating a second person in October, and by the time of Pasteur's death in 1895, about 20 000 people world wide had been treated for rabies, with fewer than 100 deaths. Despite these extraordinary figures, rabies treatment remained difficult to evaluate, for it was impossible to predict how many of these people, if untreated, would have developed the disease, or to know how many of the deaths were caused by the treatment rather than the bite.

Nevertheless the rabies work provided a fitting climax for Pasteur's remarkable career. It made headlines all over the world. The public voted with their purses, subscribing large sums for the building of a number of Pasteur Institutes, where research could continue, and making almost a cult figure out of the ageing warrior. His seventieth birthday was an international event, made even more moving by his increasing frailty and by the knowledge that for almost twenty-five years he had soldiered on with the serious physical disability of a left-sided paralysis, following an almost fatal stroke.

There can have been few scientists with a better experimental touch than Pasteur, few who can have been more passionately devoted to their craft, few whose work can have touched so many lives. Through 'pasteurisation' his name has entered the vocabulary of many languages. He provided the scientific rationale for Joseph Lister's surgical antisepsis. His work on fermentation, brewing and silk-making had important economic consequences. His advocacy of the germ theory of disease, and his researches in immunology and vaccination are pillars in the edifice of modern medical science. His speculations have not always stood the test of time, but his hunches generally paid rich dividends and he left behind a series of discoveries and techniques which have manifestly augmented our capacity to understand and control our world.

James Watson and Francis Crick with their model of the DNA molecule.

Niels Bohr (1885–1962)

Alan Turing (1912–1954)

Albert Einstein (1879–1955)

ALBERT EINSTEIN:
Questioning Space and Time
Clive Kilmister

Albert Einstein (1879–1955) is a tremendous enigma. He is, without doubt, the most famous scientist of the present century. Yet this is strange, for Einstein was never a man who sought the limelight; he was a shy, diffident and quite solitary man. Moreover, the theory for which he is best known, relativity, is one which makes very little everyday sense and has little ordinary application or relevance. One of Einstein's towering achievements was to demonstrate that Newton's laws do not always hold good. For two centuries the work of Newton had been held up to show that science was progressive. Once established, its truths were built upon but not destroyed. Einstein demonstrated that all scientific truths (as well as the relations of Nature) are 'relative'; and that (as Sir Karl Popper would put it) science works not by the establishment of greater and greater truths, but by the elimination of more and more theories that are false.

Perhaps the ideas of only four great scientists have reached the subject level of ordinary conversation during their lifetime. Two, Galileo and Darwin, clashed with contemporary theology. But the other two, Newton and Einstein, present something of a mystery. Here are two men working nearly 250 years apart in much the same field as each other. Abstract and remote from everyday experience, both captured the popular imagination of their times.

When Einstein was born in 1879 physics was entering a decade in which the idea grew up that the subject was ending. The major areas were understood, this view claimed, and little remained but to clear up one or two 'small problems'. Einstein was instrumental in changing this situation in no fewer than three fields, those of electromagnetism, quantum mechanics and gravitation. His family situation – his father and uncle ran a firm constructing electrical equipment in Munich – provided a certain scientific environment. The firm failed in 1894 and the pair began again in northern Italy. The young Einstein's education continued in

Switzerland and he entered the Zurich ETH (the noted technical university) at seventeen to read Physics, and in 1901 he became a Swiss citizen.

He left the ETH with a degree and a young wife (a fellow student) and worked as a clerk in the Patent Office in Berne; this augmented the ideas he had absorbed from the family business about electromagnetism. In 1905 he published three scientific papers in three quite different fields, all of fundamental importance. One of these, with which I shall not be concerned here, was on Brownian motion, the random zig-zag movement of very light particles scattered in a fluid, when observed under the microscope. By Einstein's time it was realised that this was caused by molecular bombardment, but he gave a complete and definitive mathematical description of the phenomenon. The other two papers dealt with topics which were to occupy Einstein for the rest of his life, relativity and quantum theory.

The paper on what is now known as the special theory of relativity was called *On the electrodynamics of moving media*. The nineteenth century had seen a very remarkable unification in physics between electrical and magnetic phenomena, each of which had its own scale of measurement. There was, by Einstein's time, a single theory, the work of many hands but owing most to James Clerk Maxwell's elegant formulation encompassing what had been thought of as two separate disciplines. His unified description of these two and the connections between them involved a number by which the quantities measured in magnetic units had to be multiplied to obtain their values in electric units. The numerical value of this conversion factor was that of the speed of light, usually written c, and having the value 300 million metres per second very nearly. Could this remarkable coincidence be explained? Maxwell's equations had solutions describing a wave disturbance and the speed of these waves turned out to be c. So was born the 'electromagnetic theory of light' in which light was 'explained' as a wave disturbance governed by Maxwell's equations and moving with exactly the speed of the conversion factor, c.

This was one of those fields in which 'small problems' were outstanding. These all related to motion. One of them, which I will call the uniform motion problem, was this: it was accepted in mechanics that an observer at rest and one in uniform motion view phenomena as happening under the same sets of forces. The uniformity of the motion is important: your cup of coffee in the restaurant car behaves exactly as at home so long as the train moves at constant speed in a straight line; but if it suddenly stops for a red signal the behaviour of the coffee will inform you of this. Yet this equivalence of rest and uniform motion did not seem to apply exactly to electromagnetic forces. A charge at rest is acted on by an electric field but not by a magnetic one. When the charge is moving uniformly it is recognised as part of a current and so there is a magnetic force as well. Of course, electric phenomena in motion are crucial in electric machinery, but pragmatic electrical engineers had not been held up by this basic fault in the theory for the difficulties were not such that judicious trial and error could not overcome them.

Another problem was inherent in the delightfully simple result that the speed of light was merely a conversion factor. The problem that this raises can best be seen by contrasting the situation with that of sound. This is also a wave motion, though described by simpler equations than Maxwell's, but for sound the speed of transmission is measured relative to the still air. A steady wind will carry the sound along faster. The electromagnetic statement says nothing about the medium of transmission or about whether the source of light is moving. Shouldn't the light from car headlights be moving faster because of the speed of the car? Not so, both from experiment and from the fact that the speed is specified by the set of equations alone, as the conversion factor c, and there is no way of taking account of the speed of the car. On the other hand, suppose that the light is moving through water. It then travels at about two-thirds of its speed in air. But if the water is moving forwards (the experiment is usually done with water moving through a pipe) it pulls the light along with it but only at just over half the speed of the water. Why is this? These and other 'small problems' had in common the fact that they involved speeds comparable to that of light.

Einstein's 1905 paper tackled these problems from a totally new point of view by asking a fundamental question that had been ignored before: how does an observer find the time at which a distant event takes place? He finds the time of local events by the use of a clock. But a distant event requires him to have information about its occurrence and this means sending the information in a signal. Different signals travel at different speeds, as one sees at once if one watches a pile-driver from a distance: the visual signal arrives sooner than the aural. So, one has to make allowances for the time of travel of the signal, and this was well understood, at a practical level. But Einstein's answer to the question was at a more fundamental level, for he held that the observer was free to *define* the time of a distant event in an obvious way as the average of the time of sending and of receiving a suitable signal between himself and the event. If I send a signal at one o'clock and receive it reflected back at three o'clock, the reflection of the signal is an event to which I can attach the time two o'clock and this is no more than saying that the signal moves equally fast on its out and back journeys.

The numerical rule is quite unsurprising. It is the emphasis on its use as a definition which is so original and carries with it the implication that there is no 'right' time of the event which the observer at rest and the one in uniform motion, in the uniform motion problem, might try to measure. Rather, each is carrying out the measurements involved in the definition of the time and it should not be wondered at if the two observers in the uniform motion problem finish with different values of time for the same event. The difference depends on the speed of the moving observer, or, to be more precise, on the square of this speed, a fact which will be important later.

But what is a *suitable* signal? The difference between the time allotted by a stationary observer and one in uniform motion depends also on the speed of the signal, and so Einstein's specification is not complete until we have chosen a particular signal. When the consequences of the definition are worked out, it transpires that the definition of the time of a distant event has two strange consequences. The signal speed will not automatically depend on the speed of its source; and it will be the greatest possible speed. The first of these consequences is the very same strange property discovered experimentally about the speed of light. By requiring the definitions to be by light signals (or, what is physically equivalent, radio waves), the definition also removes the problem about the strangeness of the speed of light and, it turns out, all the other 'small problems' mentioned above, as well as fitting in with the observed fact that no faster signals than light have been found. So the uniform motion problem is solved for electromagnetism, not by any electromagnetic cleverness, but by attending carefully to what is involved in time measurements.

This explanation is a wonderful simplification. But, as is usual in science (and elsewhere), a price has to be paid. That two observers give different times for a distant event might perhaps be regarded only as a matter of conventions being different. After all, clocks have to be synchronised, so perhaps some unexpected effect of that kind is at work. But the matter is more serious, as can be seen by looking at the case when the two observers also allot time to some second event. The difference of the times measured by each observer for the two events gives his measurement of the lapse of time between events; and not only do these lapses turn out to be different for the two observers but this difference is held by Einstein to be a result of the fact that no meaning can be given to the 'absolute' lapse of time.

This concentration on directly measured quantities like times of emission and reception of signals chimes well with the general ideas, beginning to be in vogue by 1905, of restricting theoretical discussions to observed quantities alone. Ernst Mach, for example, considered talk of atoms meaningless, as they are not directly perceived. Such a restriction is known as positivism, and people have suggested that Einstein was, at this time at least, a positivist. Such an idea fits in very badly with his later work and also with his expressed ideas and rests on a misinterpretation of what he said in the 1905 paper. The nature of this can be seen by considering the so-called 'twin paradox' devised by Einstein though it is not strictly a paradox. One twin remains at a space-station on Earth while his brother travels to a great distance at a high speed, then reverses and returns. One twin is travelling, one at rest, so they will differ about the times they assign to distant events. In particular, the times assigned to events in the moving space station by the fixed twin will differ from the local times of the moving twin. Since the difference depends on the square of the speed, the discrepancy produced on the outward journey will not be cancelled but doubled by the return journey. It is therefore an unavoidable consequence that the twins will be of different ages at their reunion. In fact the absence of any meaning in the idea of an absolute lapse of time on the journey shows up in this particular thought-experiment by the travelling twin being *younger* at the end than his stay-at-home brother. This seems unacceptable to those people who insist on relying on a 'commonsense view' based on experience of slow journeys to moderate distances but it is a small price to pay for the outstanding simplifications that Einstein's ideas brought. The real value of the experiment is that it shows that the refusal to talk about an absolute lapse of time between distant events is not a piece of positivistic pedantry but a simple statement about physical reality.

Einstein's 1905 papers made him a rising star in the scientific firmament and by 1909 he was elected to a chair in the University of Zurich, to one in Prague in the following year, to his former college, the ETH, in 1912 and then to the peak of German physics at the time, Berlin, in 1913. When the war broke out in

1914 his wife and their two sons returned to Zurich and the marriage ended in divorce in 1919. Einstein was always a pacifist and, unlike many German intellectuals, he hoped for a German defeat. But during his time in Berlin he was also struggling with the next stage of development of his relativity theory, which proved unexpectedly difficult. His first step, in 1905, had achieved the required effect of solving the uniform motion problem for Maxwell's electrodynamics. Since the uniform motion problem arises from the situation in mechanics where rest and uniform motion are indistinguishable, we can say that Einstein's 1905 paper had made Maxwell's electrodynamics consistent with mechanics, not exactly with the mechanics of Newton, but with a theory with slight modifications that showed up at high speeds.

What were these slight modifications? They were in Newton's laws of motion. In Newtonian mechanics the mass of the body is the 'quantity of matter' in it, and the body's inertia is measured by the rule that the force needed to produce a given acceleration is the product of that acceleration and the mass. Einstein showed that a formulation of mechanics which agreed with Newton's for slow speeds but which took account of the strange properties of the speed of light was possible only if the Newtonian idea of mass were slightly changed. In fact the mass of the body had to be supposed to increase with its speed by an amount which is roughly proportional to the energy of its motion. This led Einstein, in an inspired guess, to consider mass and any type of energy as equivalent concepts, distinguished only by the difference in the units conventionally chosen to describe mass and energy. The energy (E) equivalent to a given mass (M) was to be found by multiplying the mass by the square of the speed of light, $E = Mc^2$. If, then, a small amount of mass could be 'converted' into energy, this large conversion factor would ensure that a large amount of energy would be produced. But it would be overstating the case to see this as a prediction of either *atomic power* or of the atomic bomb, for although these do both succeed in turning mass into energy, they do it as part of a more complicated nuclear reaction and so depend on the much more detailed knowledge of how such a conversion could be organised; knowledge which rests on the specific facts of atomic structure.

The unification of electromagnetism with mechanics deals with the motion produced by forces. But matter itself produces forces which act on other matter – gravity. All other forces in Newtonian mechanics act only locally, as when I push something. Their action can be directly perceived; so such a force as gravity, acting at a distance, was something of an embarrassment to Newton, who tried to disarm criticism about its mysterious mechanism by his haughty 'We do not make hypotheses'. The advent of Einstein's 1905 paper, now described as special relativity to distinguish it from the general theory which includes gravity, changes the nature of the embarrassment a little, for it was not just that gravity acts at a distance but also that it does so, in Newton's theory, instantaneously. So, by moving a heavy weight here, I instantly change the gravitational forces across the road by the (tiny) amount which is the contribution of the weight. If there is a sensitive detector of gravitational force over there, this will serve as a signal whose journey time is nil. This is inconsistent with special relativity in which the speed of light is the greatest attainable.

So gravitational theory needed to be altered. But the nature of this change was not clear, the more so as Newton's theory is a very good one indeed, in the sense that its predictions of the orbits of the planets round the Sun are very accurate. It predicts that a single planet moving round the Sun will do so in an ellipse, as Kepler had observed. When one takes into account the much smaller gravitational pulls of the planets on each other, which cause these ellipses to be disturbed, these disturbances are also just as observed. By the beginning of the twentieth century, apart from a small but persistent anomaly in the orbit of Mercury, the agreement was total: far better than in almost any other scientific theory. Even the Mercury discrepancy was minute. The whole orbit of Mercury rotates about the Sun because of the disturbances of the other planets. But this annual rotation was slightly more than expected. The extra angle turned through annually was no more than that subtended at the eye of someone observing a ten-pence piece at a distance of ten kilometres.

Since special relativity showed that effects cannot follow instantaneously, Einstein tried many different formulations of gravitational theory but rejected them all, sometimes for good reasons, sometimes for bad. But at last, in 1915, and with the help of his 'friend, the mathematician, Marcel Grossmann', the definitive gravitational theory, the general theory of relativity, took shape. This was not the first debt owed by Einstein to Grossmann, for it was revision from Grossmann's careful notes when they were fellow students at the ETH that saved his original but naughty friend from being ploughed.

The logical structure of the general theory has an odd resemblance to that of the special. In each case a practical consideration is deepened by Einstein into a fundamental conceptual change. For special relativity, the need to signal and to take account of the time of flight is converted into the non-existence of an absolute lapse of time. For general relativity, already in Newtonian theory, the practical consideration is that the gravitational force acting on a body is equal to its mass multiplied by the gravitational field. This force produces an acceleration in the body by Newton's Law:

$$(\text{Mass}) \times (\text{Acceleration}) = \text{Force} = (\text{Mass}) \times (\text{Gravitational field}).$$

So acceleration = Gravitational field, and therefore if two bodies of different masses are at one place in the gravitational field, they will experience different forces but these forces will be such as to produce exactly the same accelerations. This opens up the possibility of a different treatment of gravity. Since all bodies accelerate equally in my vicinity, I can let myself accelerate with them. Around me I shall then find no gravitational field. This is the phenomenon of weightlessness, which we see illustrated by television pictures of astronauts. They, and the space-ship, are both freely-falling. Locally, then, gravity can be abolished, and the problem of describing it splits into two problems. Firstly, the distant field has still to be described, since it was probably not the same as the local one and so has not been abolished by changing to a freely falling observer. Secondly, the mathematical work involved in the change has not been done, although the physical idea is clear. This second problem is a considerable extension of the uniform motion problem of the special theory.

There is an additional restriction: the new motion problem and the new description of the distant field must be looked at bearing in mind the constancy and maximal quality of the speed of light and all the other considerations of special relativity. It was here that Grossmann was able to help; he drew Einstein's attention to the extensive work of the Italian differential geometers of the latter part of the nineteenth century, work which, with a minor change, surprisingly fitted the two gravity problems like a glove.

It is at this point that the practical considerations were again deepened by Einstein into more fundamental conceptual ones. Before 1905 the roles of space and time in physics had remained those obscurely implied by Newton in 1687 at the beginning of the *Principia* and subsequently codified by Kant in his *Critique of Pure Reason* in 1781. Kant saw them as the framework into which experience (and experiment) was to be fitted. They had to be given before experience became possible, so it was impossible to discover them by experience. In the 1905 paper the position on time was substantially modified. No longer was there an absolute 'strip of time', a sort of tape-measure along which all events could be laid out. The construction of the multiplicity of individual time-measurements which replaced it was by means of events in space, but the Kantian 'given space' was retained. The next step taken by Einstein in 1915 was to notice that the various accelerating observers in gravitational theory could all set up, locally at least, freely falling systems of space and time measurements just as in special relativity. If there is a gravitational field present, however, the different observers will be accelerating at different rates, and the effect of this will be that these systems of time and space measurements will not mesh together quite correctly where they overlap. But, of course, Kant's already given space would have ensured a correct fitting, so Kant's notion of space has to be rejected just as the idea of absolute time had to be. Instead of being given beforehand, space and time have to be seen as constructs made from the pattern of events and, moreover, the spatial construct turns out to be of a slightly different kind from that envisaged by Kant. This difference, resulting from the failure to fit together snugly, is called curvature. Curvature is produced by the field of the gravitating masses and it is this, rather than forces, that prevents particles moving in straight lines. The differential geometry explained to Einstein by Grossmann shows how to define straightest possible curves, called geodesics, which are paths of moving objects. When the geodesics round a single gravitating mass are calculated, one finds not an elliptic orbit, but a slow rotation of the ellipse. For the Sun/Mercury system the rotation is just equal to the observed discrepancy from the Newtonian theory, and so the motion of Mercury is fully explained.

Berlin was responsible for more than general relativity in Einstein's life. When he arrived in 1914 he became fully aware of German anti-semitism in a form he had not felt in his youth. This made him feel his Jewishness more and in due course led him to a rather idealised Zionism, though the spoiling of this by the 'spell of nationalism' was his constant fear. His relations with the Zionist establishment were stormy; they did not relish his public advice in 1929, after the anti-Jewish riots in Hebron, that: 'The first and most important necessity is the creation of a *modus vivendi* with the Arab people . . . Unless we find the way to honest cooperation and honest dealings with the Arabs, we have not learned anything on our way of two thousand years' suffering and deserve the fate that is in store for us'.

General relativity brought universal fame to Einstein, and not only amongst scientists. He travelled and lectured widely, and was awarded the Nobel prize in 1922. He had, by now, married again to a widowed cousin who had cared for him during a convalescence in 1919. In 1932 he was in California and so, fortunately, not in Germany when Hitler came to power. He never returned. He was appointed Professor at the Princeton Institute of Advanced Study in 1933 and settled down to a life of exile, one in which his English was always to have an extreme German accent. He was an approachable and kind man, and many young scientists had letters of encouragement in work on which they had sought his advice. His wife died in 1936 but he continued his work for another nineteen years, relaxing only by sailing or playing his violin. In 1952 he was invited to become President of Israel but quickly declined because he felt that it needed understanding of human relations and this he did not have. Indeed, from his youth onwards he had always been a very lone person.

A great deal of the last forty years of Einstein's life was taken up by a heroic assault on a general problem thrown up by his very successes in the two relativity theories. Surely, he felt, by some generalisation and deepening of the subtle differential geometry which had been so instrumental in formulating these theories, one could achieve a further unification of electromagnetism, gravitation and any other forces that might occur in physics? He tried many ways of doing this; but his investigations had to be led not by experiment, but by his critical and aesthetic senses, for there were no experimental results demanding such a unification, no measurements showing connections between electricity and gravitation. It all led to nothing and in the sixties it looked as if the whole effort had died with Einstein. Only in the last ten years has it all started again, under the heading of Grand Unification Theories, in a much more general way than Einstein envisaged, and led by the hope of explaining all the manifold experimental results about elementary particles and the structure of matter which have been acquired by the use of particle accelerators. So what had looked like Einstein being led down a blind alley had become Einstein trying to do something for which the time was not yet ripe, but giving important leads to be picked up later.

Another and faster growing branch of physics, quantum mechanics, to which Einstein contributed in 1905, occupied him at times all his life. He worked in the so-called 'old quantum theory'

had evolved, and thus by the language of visual experience. The difference was, however, crucial. Instead of attempting to des-cribe a visualisable world, which in the wake of the quantum phenomena had become a hopeless task, Bohr had only to des-cribe the phenomena as unambiguously as possible, given the language and concepts of visual imagery. If some of the conclu-sions of classical theory had to be rejected or overridden that was just too bad. It merely reflected the fact that the separation of the experiences concerned into subject and object could not be made in such a way as to allow the language of the theory to be consis-tently applied. Since, however, the conceptual foundations of this language were forced upon us, the framework and fundamental concepts of the classical theory, in which they were expressed, had to be retained. Hence the contradictions and inconsistencies of his atomic theory.

In the 1920s, as the failure of the new atomic model to predict the behaviour of anything other than the hydrogen atom became increasingly apparent, Bohr's own scientific efforts were eclipsed by those of two of his young collaborators. Wolfgang Pauli had come independently to the conclusion that subatomic phenomena were not visualisable and could not be described consistently by means of the visually based classical concepts. But unlike Bohr he did not see the language of classical physics as a constraint, and while Bohr sought to work within this language and to limit and refine the paradoxes to which this led, Pauli sought to remove the paradoxes by breaking away from the classical theory altogether. Werner Heisenberg meanwhile adopted the tactic of putting aside all such fundamental issues and trying to establish first a theory, however incomprehensible, that at least got the right results. When between 1925 and 1927 a new 'quantum mechanical' theory of physics was finally enunciated, it was largely as a result of Pauli's direction and Heisenberg's efforts. Visual concepts such as that of an orbiting electron, which Bohr had sought to retain, despite its admitted inadequacy, were unceremoniously dumped. Throughout the period in which Heisenberg, Pauli and others were developing the new quantum mechanics, however, Bohr remained the guiding figure and undisputed authority on the subject area. And, after Pauli's efforts to find a new set of non-visual conceptual foundations for Heisenberg's quantum mech-anical formalism had failed, it was indeed Bohr's philosophy that came to direct and dominate the world view associated with the new physics.

Quantum mechanics provided a set of mathematical equations

in terms of which subatomic phenomena could be described and predicted without contradiction. Since it failed to replace the language of classical physics, however, the paradoxes associated with that language, such as that of wave and particle behaviour continued to arise. In some circumstances, the (correctly predicted) behaviour of light and matter could be described only by means of particles, and in others it could be described only by means of waves.

The paradoxes remained. But it was possible with quantum mechanics to predict when and to what extent they would arise, and to relate them to the combination of properties one sought to observe and describe. Early in 1927 Heisenberg expressed these relations in terms of uncertainties in the classical description of a particle. He showed, for example, that one could if one wished observe and describe a particle's position, or its momentum. But if one sought to determine both together then the theory predicted and experience confirmed that one could get only approximate or 'uncertain' results. This conclusion, expressed in Heisenberg's 'Uncertainty Principle', was important. But in Bohr's view it glossed over the most significant feature of the new physics, and later in the same year he presented his own view of the situation in terms of a 'Principle of Complementarity'.

What Bohr did was to draw on his familiar philosophical considerations to connect the two key issues of visualisation and observation, while at the same time giving them an empirical, scientific foundation in the quantum phenomenon.

Whereas Heisenberg had equated the limits of observation of classically defined properties with the limits of their definition as given by the quantum mechanical formalism, Bohr focused on the differences between observation and definition. In classical physics it had been assumed that one could define the properties of a physical entity independent of its observer and of the process of observation. Bohr had long treated this assumption as philosophically unsound, but he now argued that it could be rejected on physical grounds. Specifically, for any process of interaction such as that between the observer and the observed entity the quantum theory required and experiment confirmed that the interaction must take place in finite quanta, the overall size of the interaction being governed by the quantum constant. One simply could not observe a particle, say, as classical theory had assumed, without disturbing it by more than an infinitesimal amount.

This in itself would have been no problem, had one known the full details of the interaction involved. But the quantum theory stipulated only the overall size of the interaction, and theory and experiment were at one in asserting the impossibility of determining how it was made up, an impossibility that could also be expressed philosophically in terms of Bohr's old favourite problem. To determine the details of the interaction one would have to observe it, and that observation would itself require an interaction the details of which were not known. To determine these details one would have to make another observation, and so on. The consequence of this impossibility according to Bohr was that, unlike in classical physics, one could not define and observe a system at the same time. In order to define a system precisely, one would have to eliminate all external disturbances, but that would preclude any act of observation. To observe it one would have somehow to interact with it, and since the details of the interaction could not be known any definition of the system would be then impossible.

In Bohr's terminology the definition and observation of a system were complementary. Both were necessary to any physical theory, but they were in fact mutually incompatible, and only if one accepted a measure of uncertainty or approximation could one talk of both together. From this primary complementarity, he derived other complementarities, including those which had already found expression in Heisenberg's uncertainty principle, and that between the wave and particle interpretations of light and matter.

Bohr's contributions to physical theory did not stop with quantum mechanics, nor did they stop in the 1920s. During the 1930s he propounded the theory of the compound nucleus and, on the eve of the Second World War, that of nuclear fission. He also continued to refine and restate his complementarity view of physics in an effort to convince those physicists who were not won over to the new orthodoxy of 'Copenhagenism'. In particular he sought to win over his friend Einstein, who persisted in hoping for a complete theory that would allow all physical phenomena to be explained within what was essentially a classical framework. For Bohr, whose whole philosophy revolved around issues of language and communication, the failure to convince someone of his position could only be interpreted as a failure to communicate this position adequately. Over and over again he would argue with Einstein, or with Einstein's image when he was not there, or with his ghost after he died, ever seeking a clearer way of putting forward his argument.

It is impossible, however, to understand Bohr's place in history solely in terms of his own physics and philosophy. From the 1930s he played a leading role in international science politics, masterminding the relocation of displaced Jewish physicists before the war and working actively with the Danish resistance, and later with the Allied atomic bomb project, during it. Once the atomic bomb had been developed and as soon, indeed, as the threat that Hitler might develop and use it had been removed, Bohr's was the leading voice in the call for international collaboration and control as a way of securing world peace and trying to ensure that the bomb was never used. In some of these activities Bohr was successful. In others, such as the control of atomic weapons, he was not. But throughout them all he displayed the same qualities he did as a scientist, and in particular the qualities of humanity. Had Bohr been a remote ivory-tower scientist his role in history would still have been considerable, but he would have had nothing like the impact he did have. His was a social and sociable character, marked by a strong sense of humour and a sense of adventure, by childlike fun and fatherlike concern. He attracted people towards him and in so doing he attracted them towards his ideas, not as a gospel truth but as a source of inspiration. The ideas themselves took on a tremendous importance in the history of science, and the values accompanying them also marked a generation of scientists.

ALAN TURING:
The Mind and the Machine
Andrew P. Hodges

Conflict between societies has always been a great stimulus to invention. As with other technological advances, the rapid development of the computer began during time of war. Alan Turing (1912–1954) was amongst those academics who were thrown together during World War II in the national interest. In the thirties, Turing had shown himself to be an original and highly talented mathematician. During the war he applied his ideas on the nature of thought and logic to help create the first embryonic computers, which ultimately unlocked essential secrets to the allies. From such primitive machines he envisioned a computer industry and machines capable of intelligent thought, and he devised the definitive test for determining artificial intelligence.

When the history of science reaches Alan Turing, it hardly feels like history at all. For at the centre of his life is the computer, and everything that is or could be done with it. Yet Alan Turing lives only in history; he died in 1954. It has taken a generation to catch up with him.

There was nothing very modern about Alan Turing's origins. In fact they were of a distinctly old-fashioned kind. His father was an official in the Indian Civil Service, and his education was in traditional English preparatory and public schools which paid little attention to science. He emerged as a quiet, firm, intensely individual, and isolated person.

He was also a very unusual character, full of all sorts of quirks sometimes endearing and sometimes irritating. He was not a person who fitted in – either socially, or scientifically. He would certainly not fit into the picture of scientists busily making observations and correlating the results. Turing himself wrote:

The popular view that scientists proceed inexorably from well-established fact to well-established fact, never being influenced by any unproved conjecture, is quite mistaken. Provided it is made clear which are proved facts and which are conjectures, no harm can result. Conjectures are of great importance since they suggest useful lines of research.

He could have been speaking about quantum mechanics and relativity, which had demanded astonishing conceptual audacity. But he was also referring to the way he himself had dared to make new pictures, new connections. To study Alan Turing is to study the power of the scientific imagination.

So let us come to the computer, Turing's great central idea. Usually the idea of the computer is introduced thus: first you are presented with a machine you can see and touch (hardware). Then programs (software) are written for it. Each program, when fed in, makes the computer behave in a different way so that it can do a different job. On top of this there is daring talk about computers showing a form of intelligence. Histories of the computer tend to be similar: people build the machines, then write programs for them, then speculate about intelligence.

But if we follow Alan Turing's story we encounter these ideas in the reverse order. First Turing thought about mind and intelligence. Then he came to thinking of all the jobs that a machine could do; then arrived at the idea that all these jobs could be given to a single machine. Finally, he realised that such a machine could actually be built, and embarked on doing just that – on building the first electronic computer. And for him, the computer was primarily a laboratory for finding out about thinking. Each step is a different and fascinating story in its own right. We shall follow them in turn.

There is a classical problem which goes to the heart of scientific inquiry. Science tries to find laws of regular behaviour, and the laws of physics can give predictions of fantastic precision. In practice we can apply such laws only to rather special systems like the motion of the planets; but there is nothing in physics to say why the same laws should not apply to every atom in our bodies – and brains. If indeed physical laws do govern all matter in our brains, why do our minds have a sensation of choice and freedom? Anyone really serious about science must at some stage be deeply perplexed by this question, along with the scientists and philosophers of every century.

The advent of quantum mechanics in the 1920s, through the work of Einstein, Bohr and others, had for the first time suggested a way out of the apparent paradox. It could be claimed that physics was not deterministic after all; that nothing told the radioactive atom when to decay. The astronomer and physicist Sir Arthur Eddington grasped keenly at this idea and his book *The Nature of the Physical World* (1928) held that the mind could manipulate the undetermined movements of matter so as to enforce its own free will. Turing read Eddington's famous work at school and formed his own view, expressed in a letter of about 1932, when he was twenty: 'We have a will which is able to determine the action of the atoms probably in a small portion of the brain . . .' Turing's interest in this question was greatly sharpened for emotional reasons. At this stage he very much wanted to believe in mind independent of matter, for he wanted to believe in mind surviving death. His friend Christopher Morcom had died at school in 1930. The letter with this theory of mind and brain was written to the boy's mother.

The obvious area for Turing to study, to understand these questions was physics. Such a choice would have been natural for any mathematically minded young scientist in the 1930s. But instead, after trying various different areas of mathematics, he settled on the field of mathematical logic. On the face of it this was a quite unrelated area. You will have to bear with some explanation until the connection that Turing made becomes visible.

From the 1880s onwards an attempt had been made to break down mathematics into its most primitive elements in order to found mathematics on a solid basis of unquestionable truth. Bertrand Russell was perhaps the best-known person involved. But Russell showed it was very hard to find these solid foundations. The trouble comes because mathematics involves statements about infinitely many objects at once: such as 'every integer can be written as the sum of four squares'. Attempts to nail down the use of such infinite sets led to paradoxes which could not be satisfactorily resolved.

David Hilbert, the great German mathematician of the early twentieth century, looked at these questions from a new angle. Hilbert said that any proposed system for mathematical foundations, such as the one Russell had tried to work out, should satisfy three requirements. It should be consistent. This meant that one would never follow rules of the system and end up by showing $2+2=5$. The system should be complete. This meant that any true statement could be proved by using the rules. And it should be decidable. This meant that there should exist some definite method which could be applied to any given mathematical assertion, at least in principle, and would be sure to decide whether that assertion were true or false. Hilbert thought these were perfectly reasonable demands, but in 1931 Gödel showed that Hilbert was wrong. Mathematics could not be both consistent and complete. Gödel achieved this by constructing a mathematical assertion which effectively said 'This statement cannot be proved'. Now, either this assertion could be proved, in which case mathematics would have an inconsistency; or else it could not – in which case it would be true but unprovable, and mathematics would be incomplete. It was an astonishing discovery.

Turing learnt of this in 1935, and learnt that no-one had yet been able to deal with the remaining question of whether mathematics was decidable. The difficulty lay in giving a definition to the idea of 'definite method' or 'rule'. What did one mean by such an expression? The whole point of such a 'method' was that it could be applied mechanically, without requiring any further human choice or invention or ingenuity. People had in fact used the phrase 'mechanical process' and it was on this idea that Turing seized. For what is a 'mechanical process' – but something that can be done by a machine.

This is where Turing dared to think of a new picture: a picture of machines that could do mathematics. The machines Turing devised were machines in a thought-experiment, not machines that needed actually to be built. They were imaginary machines that could read mathematical symbols and work away to produce answers to questions. What he managed to do was to break down all mathematical work into little atoms of logical operations, and then imagine machines doing them.

The picture Turing gave was this. First, a paper tape in which symbols can be written. Then, a machine which can look at just one symbol at a time. The machine can be in one of a finite number of different states. What it does – how it behaves – depends only on what state it is in, and the symbol it is looking at. And all it can do is: to move one place to the left or the right on the tape, to write a symbol on the tape, and to change its state to another (specified) state.

Such a machine is called a Turing machine. Each machine has a definite rule for how it will behave in each state, for each symbol it sees. Sometimes people speak of 'the' Turing machine, but this is misleading because there are infinitely many different Turing machines, each one with a different rule about how it is to behave. You can specify a Turing machine precisely by a table which says what it will do in any state, when it sees any symbol. Given the table you know everything that matters about the machine. Now Turing claimed that anything you might call 'a definite method' could be thought of as a Turing machine.

Turing gave several arguments for this remarkable claim. One showed how adding, multiplying, and all other mathematical operations could be built up from the little atoms of operations that the Turing machines are allowed to perform. But the real power and interest of Turing's claim really lies in the way he went outside mathematics to justify it. He imagined a person carrying out something that one would call a 'definite method'. He argued that such a person could at any stage leave a complete note of what was to be done next, in such a way that someone else could take over. Then he showed that this is equivalent to specifying the 'table of behaviour' of a Turing machine.

But he wanted to go even further, and here at last we see his interest in the nature of the mind manifesting itself. For Turing argued that Turing machines could represent anything the mind did in carrying out a 'method', even if that method were not described explicitly at every stage. His argument was that the mind could only be in one of a finite number of 'mental states', and that the 'method' could be described as the way that those mental states would change when faced by the symbols. He argued that mental states could then be represented by states of a Turing machine. So he arrived at the claim that any mental process can be represented by a Turing machine.

By employing this definition of 'definite method' he answered the outstanding question in mathematical logic. It appeared in the now famous paper he wrote in 1936, explaining all these ideas, when he was just twenty-four. For he showed that no Turing Machine could do what Hilbert asked, namely test all mathematical assertions for their truth or falsity. Turing showed this by supposing that such a machine could exist, and then setting it to work on an assertion about its own behaviour – an idea rather similar to Gödel's construction of a statement referring to its own provability. This he showed led to a logical impossibility, showing that no such machine could possibly exist. It needs an infinite supply of new ideas to solve mathematical problems. But while showing that there was something that machines could not do, he found an idea which was to play a most important part in exploring what machines could do: the idea of the universal machine.

The concept of the Turing machine led Turing to the idea of 'universal machine'. The argument is this: the business of looking up the entries in a 'table of behaviour' is a mechanical process – so a Turing machine can do it! In fact a particular kind of Turing machine, the universal Turing machine, can be designed to read the table of behaviour of any other Turing machine, and then do what that other Turing machine would have done. Thus, it is not necessary to build more than one machine in order to see the behaviour of any Turing machine.

This beautiful observation is an example of the elegance and economy that makes mathematicians enjoy mathematics – it is a wonderful thing to see how once a theory has been set up, it will generate new ideas. But statements in mathematics that such-and-such can in principle be done, are not always useful in practice. Turing's published description of the idea of the universal machine did not discuss whether such an idea could be given a practical shape. We do not know exactly how much attention he gave to this idea at first. But he certainly took up an interest in practical machinery over the next few years.

You will see that the universal machine embodies the idea of the computer. The computer is designed to read a program of instructions and carry it out. The machine itself remains untouched – we need never open the box, only feed different programs into its memory. One machine suffices for all the different tasks. The 'methods', the Turing machines, are the different possible programs. Reading Turing's paper it is hard to remember that the computer did not exist in 1936 and that Turing's work was of the imagination.

Everything changed through the outbreak of the Second World War. The war has often been called the physicists' war, but it was the mathematicians' war too, in a way that was kept secret for thirty years. Alan Turing was at the heart of it: he became the chief British cryptanalyst. The problem that faced him was the now-famous Enigma machine, with its complicated permutations. This cipher machine was used throughout the German war effort – army, navy, air force, the lot – and it held the key to its communications. It was not too difficult to find out all about the machine – but the machine could be set up in thousands of millions of ways, each one creating a different encipherment process, and the job of the cryptanalyst was to work out which one had been used for each message.

Polish mathematicians had pioneered some clever analysis of the Enigma-enciphered messages, but by 1939 the system had been improved and their methods had become inadequate. Turing and another Cambridge mathematician, W. G. Welch-man, by building on the Poles' ideas, made the first breakthrough. The logical trick that Welchman and Turing found could be embodied in the electrical circuits of a special new machine which automated a crucial part of the cipher-breaking process. This machine could be thought of as a practical Turing machine. It did not actually have a paper tape but the essential idea was the same. For it had to be supplied with information: a small piece of cipher-text and a guess at what that text encoded. Then it would work away carrying out its definite logical method and produce an answer to whether the guess was right and if so, in what setting the Enigma had been used for the enciphering.

But in fact there was not just one moment of 'breaking the Enigma'. There were different variants of the Enigma, all used in different ways, the different methods had to be found in each case. It was a miracle that this was possible at all with the technology available at the time. Indeed, when the Enigma was used properly it was not breakable; and an improvement to the Enigma used by the German naval forces, at the beginning of 1942, robbed the Allies of decipherment for nearly a year with disastrous results. The whole thing rested on a knife-edge.

Turing's direct responsibility was for the naval Enigma, which because it included the U-boats was probably the most vital as well as the most difficult. Speed was the paramount requirement, for ciphers had to be broken in hours to make use of the U-boat orders and reports they disclosed. The result was to crush into a few years developments which might otherwise have needed decades. It was not just a question of making new machinery, but of inventing new types of logical processes to use that machinery for. People and machines were organised into an enormous system all concerned with performing definite methods with symbols – in fact, with Turing machines of incredibly practical value.

The demand for speed brought electronics into the picture in 1942, and thus Turing learnt of and used what was then the newest technology, capable of performing logical operations a thousand times quicker than the telephone exchange technology which the earlier electrical machinery used. Large and successful electronic machines were built in 1944 and Turing also gained personal experience with electronics from his top-level mission to the United States, where he was given access to the most advanced work on speech encipherment. Electronic speeds made it possible to think about building a practical form of the universal machine, and by 1945 Turing was talking of wanting to 'build a brain'.

Turing got his chance; he was installed at the National Physical Laboratory in October 1945, with the job of designing an electronic computer. In a few months he produced a detailed working plan. He described it in terms of the universal machine: by feeding different instructions into its memory (now to be done electronically) one would get it to perform any desired procedure.

At this point, we encounter one of those questions which make the history of science so fascinating – and difficult. For Turing was not quite the first with this ambition. His post at the NPL owed much to the fact that in June 1945 an American plan for an all-purpose electronic machine had been announced. The writers of this report, which included the great Hungarian-American mathematician John von Neumann, had arrived in their own way at the idea of storing instructions in an electronic memory and getting the machine to read them and carry them out, just as in the universal machine. Turing referred to the American plan in describing his own. It is not possible to say how much von Neumann was helped by knowing of Turing's 1936 paper. Turing himself made nothing of the question of who was first with the idea of the computer. What he cared about, and where he was really far ahead, was using the principle of the universal machine to see how to exploit the computer most effectively and excitingly.

You will remember how Turing's 1936 paper involved the definition of 'tables of behaviour' for Turing machines. These could now be thought of as lists of instructions – programs – for the computer to carry out. In devising particular 'tables' in 1936, Turing had already got into the frame of mind of the computer programmer.

But he also brought from his earlier work the understanding that the computer should be thought of as operating not on numbers, but on symbols, and the symbols could refer to anything whatever. It was an insight that came from his background in modern, abstract mathematics, where symbols are used freely and without any necessary connection with numbers. It was also an insight that came from his wartime work when the symbols studied were not so much numbers as texts. Turing saw the computer not as a machine to do calculations, but as a machine to process information in any way desired. His computer plan proposed applications – chess-playing, puzzle-solving – far removed from arithmetical calculation.

He also saw that the instructions to the computer were themselves just symbols, no more and no less than symbols representing numbers or other forms of information. He realised this meant that the computer could be used to process its own instructions. It would be possible to write the instructions in any way one liked, whatever was convenient for the user, and get the computer to do the drudgery of translating these instructions into the form it actually used. And thus he arrived at the idea of computer languages.

These were completely new ideas. The general idea of a machine to do all arithmetical calculations was not new. Charles Babbage, the British mathematician and scientist, had designed such a machine a hundred years before. Babbage had never been able to effect the mechanical engineering required, but his ambitions had been revived in the 1940s in a number of American-built electrical calculators. These machines, however, were designed to store numbers in one place, and then in some other place to hold a sequence of instructions on what to do with those numbers. Numbers and instructions were treated as entirely different and separate entities. Turing, however, saw from the start – indeed, he had seen it in 1936 – that numbers, data, instructions, were only different examples of logical symbols, which could all be stored in the same form as electronic pulses within a computer memory. This perception was rooted in the development of modern mathematics, and in particular the mathematical logic of Russell, Hilbert, Gödel and others of Turing's predecessors; it was with Turing's application of this perception to practical problems that modern computing begins.

Turing was also far ahead of others in visualising not just a computing machine but computer systems, computer centres, almost a computer industry, with the dynamic development of new hardware and software. It was a picture he could have gained from his wartime experience. However, even the machine Turing proposed to start this development off with (though ludicrously small, at 32 Kbytes, by present-day standards) was too ambitious for the NPL, which insisted on cutting it down. Alas, nothing was built in three years. Turing was disappointed and angry.

He was particularly frustrated by not being given any say in the electronic engineering required. Some people would say he exaggerated his competence to do this, and had an arrogant view of engineering as something anyone could pick up. On the other hand, Turing at least had the virtue of being prepared to get his hands dirty; not at all the academic theoretician. He would have liked to carry on at the pace of the war, when projects like this had been carried through with night-and-day working and he did not adjust to the peace.

Turing resigned from the National Physical Laboratory, and lost his place in the practical advance of the computer. His advanced programming ideas were lost and forgotten. If he had combined his mathematical and logical knowledge, his wartime experience of organising problem-solving efficiently, and his ability to communicate and promote ideas effectively, he could have remained the towering figure in the creation of computer science and information theory. He abandoned this role.

But he did not lose interest in the idea of getting a computer to display intelligence. For it was the problem of understanding the nature of thought that had motivated all his work. Having jumped immediately at the idea of processing instructions as well as data, Turing was also attracted to the idea that the computer could easily make changes to the program as it went along, according to its experience. He saw this as a picture of how a computer could learn, a vital ingredient of intelligence. He also explored the more ambitious idea that 'states of mind' could be thought of as states of a Turing machine, and considered ways in which to get a computer to learn without any explicit programming, by a process more like evolution. He thought that 'intelligent machinery' – what would now be called 'artificial intelligence' – would develop by a combination of these methods.

Turing was aware that talk of 'intelligent machinery' sounded silly or shocking, and rather enjoyed it, becoming a witty populariser of his ideas. His argument was essentially this: we do not know how the brain works. However, all that is important about the brain is the logical pattern of the nerves – the brain regarded as a Turing machine. That the brain is grey and squishy is irrelevant to its capacity for thinking. Now, this logical pattern – this Turing machine – could be embodied in some other physical form. But there is no need to make a machine whose connections are anything like those of the brain. For the computer is a universal machine, and so a computer with sufficient storage capacity will suffice. All we need to do is to give it the right instructions.

Opponents objected that a machine could never think in a human way; it could only coldly calculate; but Turing took a rather aggressive, cheerfully flippant, line on this and said that we should have to wait and see. He was quoted in *The Times* as saying 'Maybe a sonnet written by a machine will be better appreciated by another machine'. The playfulness disguised a seriousness about his own argument: if it was the logical pattern of the brain that mattered, this logical pattern must include all that goes on inside human brains, whether cold and calculating or not.

Turing's best-known paper on this subject, written in 1950, tried to convince the reader with the witty description of an Imitation Game, a thought-experiment designed to screen out the irrelevant factors of appearance and outward behaviour, and to concentrate on what goes on inside the head – or the machine. Turing set up the game by imagining an 'interrogator' communicating over a remote teleprinter with a 'witness' which might be either human or a computer. The interrogation would range over all subjects, which (waving a red rag at the bull as usual!) he imagined taking in witty comments on English literature. He argued that if the interrogator could not tell a machine from a human being, under the conditions of such a test, then the machine must be credited with intelligence.

This is really the same principle as examinations and 'intelligence tests', and one may feel dubious about the so-called Turing Test for similar reasons. But it is important as a thought-experiment for what sounds so dull but is so important to science – the process of model-building: deciding what things are relevant and concentrating on those, discarding the irrelevant. Turing's claim is that 'thought' or 'intelligence' is completely describable in terms of the Turing machine model. The squishiness of the brain, and its embodiment in quarrelling bipeds, he held to be inessential features. We may doubt that this is the whole story, but the Turing machine gives a new model of thought-processes which opened the way to experiment – and a vast range of practical applications.

Turing's work had not been confined to the computer. It had ranged over many other mathematical and scientific topics. So relinquishing the development of the computer did not leave him bereft of ideas. On the contrary, there were many loose ends he could profitably have taken up. What he chose, however, was to start again in 1950 with an entirely new problem – in biology.

Watson and Crick, amongst others at that time, were trying to see how genetic information was stored in the cell. Turing concerned himself with a complementary question: how can such information possibly give rise to the patterns of biological growth? How does a plant know when to put out a new leaf? The question was closely linked with the old puzzle of physical determinism: how can pattern arise out of a mere soup of chemicals? Turing found a possible mechanism by studying the equations of chemical reactions. He discovered a chemical effect analogous to mechanical instability – the idea of the last straw on the camel's back. At the moment when the last straw is piled on, the camel collapses, but it is impossible to predict whether it will fall to the left or the right. In the same way, a homogeneous chemical solution can become unstable (by being heated, perhaps) and at the critical moment loses its symmetry by singling out a direction that cannot be predicted. For example, Turing considered the problem of how a growing animal embryo suddenly loses its original perfect spherical symmetry by developing a sharp fold which goes on to become the spine. Turing also pioneered the use of the computer for simulating the very complicated effects involved. This was work heading towards a Nobel prize, and shedding new light on those old questions about physical determinism. But his research was interrupted.

For Alan Turing the scientist was also – according to the law – a criminal who was caught by the police early in 1952. He was a homosexual whose crime came to light because he resisted a petty form of blackmail arising out of an affair with a young man in Manchester. Turing refused to say he had done wrong, but he had foolishly given the police a statement and had to plead guilty.

He was given a year's probation on the condition that he submitted to hormone 'therapy' – a kind of chemical castration then coming into fashion for 'sex offenders'. More trouble lurked behind the scenes. Turing had continued to do secret work for what had now become GCHQ. The mathematicians' war, like the physicists' war, had not ended in 1945 – although it remained a lot more secret. The disclosure of his sexuality obliged him to cease such work, and his visits abroad created further anxiety for the state. None of this explains his death in June 1954, but one can safely say that it took place against a background of acute pressures. He died by eating a cyanide-poisoned apple.

It is hard to assess Turing's direct influence on the practical building of computers. Probably it was slight – like the events at the end of his life a warning that our so-called civilisation can easily wreck and waste and cover up the damage. But his influence on the theory and scope of what computers do remain immense. If anything it has grown since the 1970s. People are still thinking of new ideas to do with Turing machines; artificial intelligence is just at the beginning of practical application; meanwhile biologists have hardly yet caught up with his theories of growth. It is far too early to give any assessment of his place in the history of science.

JAMES WATSON and FRANCIS CRICK:
Discovering the Secret of Life
Edward Yoxen

Scientific progress relies on the competitive spirit. Pasteur continually battled with the German chemist, Liebig, to prove the nature of fermentation, and Priestley and Lavoisier did much the same in their race to account for the properties of air. But nowhere has this competition been more obvious than in James Watson's account of the discovery, by himself and Francis Crick, of the double-helical structure of DNA – the key genetic substance. Watson has portrayed this major development as a race for fame, involving cut-throat competition and a fair bit of double dealing. Regardless of how it came about, there is no doubt that the unravelling of DNA has been of immense importance. Once it was understood how genetic information was passed on through genes, the possibilities of changing that information became real, opening up today's new field of genetic engineering.

DNA is ubiquitous. It forms the genes of virtually every living thing on this planet. It is the substance in which the genetic programme is stored. As such it appears in every single cell, except for those few that live on without chromosomes. Development, metabolism and reproduction depend crucially upon it. We have become accustomed to say, without noticing the exaggeration any more, that DNA contains the secret of life.

Forty years ago one would scarcely have written in this way. The biological role of deoxyribonucleic acid (DNA) was still in doubt. Genes were thought to be made of protein, with DNA responsible for some very minor role in the cell nucleus. Its chemical composition was only partially understood, and its three-dimensional structure unknown, although several scientists had made guesses. After 1953 much was to change. In that year James Watson and Francis Crick proposed that DNA forms a double helix, composed of two phosphate-sugar complementary strands wrapped around each other and bonded together at intervals across the centre of the molecule. By now this simplified representation (see p. 208) is known to anyone who has done even elementary biology. It is the most common scientific symbol of our age. As a logo it has appeared on the spines, covers and title pages of countless publications. In several senses it is the quintessence of modern biology.

This elegant model suggested immediately how genes could go on copying themselves, in cell division after cell division. It also showed that genetic information could be physically embodied in the sequence of chemical bases at the centre of the molecule. In this way hereditary traits could be specified by sections of DNA. Quite how this was done was not immediately apparent in 1953, but it focused the inquiry. It gave metaphors like 'hereditary information' and 'genetic programme' drawn from the sciences of computing, cybernetics and information theory, a concrete meaning as structures within a molecule. It offered a very graphic way of posing problems in a new scientific field called molecular biology.

To write of the work of Watson and Crick on the double helix as discovery is quite reasonable but potentially misleading. They did not discover DNA as a substance. That had been done in 1869. Nor did they discover what it did in the cell nucleus. That was clarified to many biologists' satisfaction by 1944, with further evidence appearing in 1952. Nor did they discover any novel facts about its chemical composition through experimental investigation. A vital piece of information about the chemical units at the centre of the molecule was produced by Austrian emigré, Erwin Chargaff, working in the US in 1950. More data about the strands on the outside emerged from the work of two British chemists, Todd and Brown, in 1952. Nor did Watson and Crick discover how crystals of DNA diffract a beam of X-rays to form a characteristic pattern on a photographic plate: that was first done in 1938. Rather Watson and Crick subjected others' results in X-ray crystallography and biochemistry to critical scrutiny, extracted new significance from them, and proposed by reasoning and imagination alone a hypothetical model of the three-dimensional structure of DNA. Their model was immediately appealing, but it did not explain everything and it could have been wrong. Much confirmatory evidence in its favour now exists and today we take it, with a few modifications, as correct. In this sense then Watson and Crick discovered new aspects of an already known substance, which allowed then to explain how it played its vitally important role.

It is the speed with which they moved to the solution, the

mastery of the available data and of techniques in several fields, and the confidence with which they set to work on a difficult problem that makes their achievement so remarkable. This mix of indebtedness to others and competition with them is common-place in science today, so vast an enterprise has it become, and it makes an obvious contrast with other examples in this book. In the 1950s reliance on the results of previous generations was inevita-ble, but the constant sense of racing against competitors all around the world was much weaker, except for young men in a hurry, like James Watson.

So preoccupied had he been with the professional fact of competition, and with the thought that missing clues might lose the aspiring biologist valuable ground, that he made it the under-lying theme of his own iconoclastic and highly revealing book about his work with Crick on DNA, itself called *The Double Helix*. When this appeared in 1968 it scandalised the scientific world as an all-too-honest account of what he in particular, and scientists in general, were prepared to do to get a competitive edge.

He was remarkably frank about some of his friends' and col-leagues' shortcomings and the book appeared only in the teeth of opposition from some people named in it. At least one book has appeared to set the record straight. Even though some of the people given this treatment have now decided to forgive and forget, it amazed many by its harsh, lacerating tone. It has of course sold extremely well.

Watson has said that he was trying to show that the discovery of DNA was really not as simple as commentators implied, and also that he wanted, without real malice to anyone, to write a moral tale that would inspire young people at school to enter biology. In his youth he had been greatly impressed by Sinclair Lewis's novel *Arrowsmith* which concerns the struggles of a young medical researcher in his attempts to apply the newly discovered bacteriophages in treating disease; Watson had used phage thirty years later in his genetics experiments, and wanted to produce something similar. His basic message in *The Double Helix* is that doing biology today means thinking more quickly and more pre-cisely than others who had been myopic in thinking about DNA, either in believing results they ought to have doubted or in missing the significance of a valid result. Such stupidity cost others the race, as it might, Watson implies, have done for him too.

Though Crick has been scathing about his portrayal in *The Double Helix* there is no doubt that he too is proud of the changes that molecular biologists have wrought in the life sciences. Nowadays biologists tend to think more theoretically, to focus their attention on the molecular level, to use many more physico-chemical methods, and to repudiate as mystical nonsense the philosophical vitalism of earlier generations that held living matter to be something special. This comes through clearly in Crick's popular writings on the the nature of life in this and other solar systems. Life for him, as for many molecular biologists, is the processing of information in organisms of evolving complexity. DNA has now a central role in the world-view of scientific mater-ialism. That is why the discovery of the double helix was such an important event.

The substance we now call DNA was first isolated by the German scientist, Friedrich Miescher, in Tübingen in 1869. His interest was the chemistry of basic cell components. Working with pus cells from discarded bandages, he found he could produce a precipitate of some substance from the nucleus that was not a protein. Further work showed it was to be found in many other kinds of cells. He called it nuclein. He did not suppose that it had any genetic role. He thought it was just where cells stored phos-phorus. Given the lack of understanding of inheritance at the time, this was a plausible guess.

The latter half of the nineteenth century was a period of great uncertainty about the mechanism of heredity. Darwinian evolu-tionary theories suggested that species remained distinct for long periods of time, but also that within species particular characteris-tics conferred a selective advantage on individuals inheriting that trait. Yet how such characteristics were passed on remained obscure. By the 1880s better microscopes and new stains allowed biologists to identify and keep track of structures in the cell nucleus, like the rod-like bodies called chromosomes. They seemed to split and group themselves into two subsets, one on each side of the dividing cell. This suggested that the chromo-somes were the bearers of hereditary traits, but other interpreta-tions were possible. Knowing that they were made of protein and Miescher's nuclein, by then more accurately characterised as nucleic acid, was very interesting. But in itself it explained very little.

By 1920 much more evidence was available that chromosomes did indeed carry hereditary traits and that the genes, the determi-nants of specific characteristics, were strung out along them, like beads on a necklace. This was accepted as the basis of a theory of inheritance by many biologists of the day. More chemical detail was also available about the nucleic acids. The German chemist,

Albrecht Kossel, had identified a set of four chemicals – or bases – adenine, guanine, cytosine and thymine as constituent sub-units in the 1890s. His compatriot, Emil Fischer, had synthesised them from known chemical compounds, which gave information about their structure. In America the Russian emigré, Phoebus Levene, working at the Rockefeller Institute for Medical Research, had identified a sugar molecule with five carbon atoms (deoxyribose) as another component. Later he put forward a simple model of DNA structure, known as the tetranucleotide hypothesis. He envisaged DNA as a long chain in which building blocks called tetranucloetides repeated themselves monotonously. The building blocks in turn were made up of four subunits comprising each of Kossel's bases in turn. On this view DNA had a regular structure and it seemed impossible that it could specify *different* genetic traits. It might be a kind of scaffolding within the chromosome. Levene's status as a chemist and the lack of direct evidence to contradict his hypothesis meant that this remained the orthodox view until the mid-1940s.

The phenomenon that led eventually to the overthrow of this view came from bacteriology. In 1928 an English microbiologist, Fred Griffith, observed that particular strains of bacteria acquired new characteristics, which their descendants retained, when grown with an extract from another strain possessing these characteristics. One disturbing interpretation of this was that some substance was capable of passing on heritable traits. The organisms were in effect picking up new genes. This struck directly at existing ways of naming and identifying bacteria and of raising specific antisera against them in the days before antibiotics, because it implied that the differences between strains were not completely fixed. Moreover, most microbiologists believed then that bacteria did not even possess genes. Here was evidence that they did. One scientist provoked by this discovery was Oswald Avery, working at the Rockefeller Institute in New York. After many years of painstaking labour he and his colleagues showed in 1944 that the gene-like 'transforming substance' in Griffith's experiments was in fact DNA. The most likely interpretation was that bacterial transformation occurred through the transfer of a set of genes from one strain to another and that all genes, not only in bacteria, were made of DNA.

These ideas were of particular interest to a group of young American scientists, led by two refugees from Europe, Max Delbruck and Salvador Luria. They had just begun to do genetic experiments with viruses that infect bacteria, called bacteriophage, to find out how genes copied themselves. To them, although by no means to all geneticists, the question of the structure and chemical nature of the genetic material was of very great importance. One member of this group was Luria's student, James Watson.

Watson was born in 1928 in Chicago. He was a very bright child, a radio quiz show champion. He went early to the University of Chicago to read zoology, and then on to Indiana to do research in genetics. By twenty-two he already had a Ph.D. Characteristically for an American graduate student he had been coached very thoroughly. He had been taught where the interesting problems lay and shown how one might attack them. Delbrück in particular, as one of the organisers of a summer school in genetics that Watson attended, believed in trenchant criticism of one's own and other's sloppy thinking. In 1951 Watson's scientific patrons encouraged him to go to Europe to acquire new research skills. In particular he was interested in DNA. He had come to the view that the only way to discover its structure was through X-ray crystallography, in which he had not been trained. Accordingly, against the wishes of the National Foundation for Infantile Paralysis, which paid his way, he got himself invited to the Cavendish Laboratory in Cambridge. There the Medical Research Council was funding a small unit working in a Nissen hut. Its members were trying to work out how proteins like the red blood cell pigment, haemoglobin, were folded up. Their method was to work backwards from the patterns made by passing a beam of X-rays through haemoglobin crystals to a model of its structure. It was an arduous task, with vast amounts of data, which even very sophisticated mathematical processing could not subdue.

One member of this research group was Francis Crick. He was born in 1916 into a wealthy middle-class family, whose fortunes survived the demise of their Northampton shoe factory during the 1930s. He graduated in 1937 with a physics degree from University College, London and started work for a Ph.D. With the advent of war he was drafted into Naval intelligence where he was employed in finding new ways of sweeping mines. The work, on one of the precursors of operations research (OR), suited his deductive way of thinking from a few assumptions that cut through the complexity of the problem. This style of thinking runs all through his scientific papers, sometimes with spectacular results. When he was demobilised in 1946 several distinguished scientific mentors were happy to help this bright young man, now thirty years old, resume his career in some area of biology, even though his forthright manner had sometimes caused problems for slow-witted military superiors.

After two years in biophysics, reading himself into a new literature, he joined the protein crystallographers at the Cavendish. He was thus very much on the margins of biology, but had his own views about how to tackle the significant problems, which derived from his initial training. He was also developing a very specific expertise in crystallography. But progress for the group as a whole was slow. Crick found it hard to stick to his assigned piece of the puzzle and tended to interest himself in others' problems and comment on their theoretical significance. At both the personal and professional levels this is a risky research strategy, leaving the tedium of data collection to others, whilst you produce the interpretation that cracks the basic problem. It does not suit the apprentice researcher. By 1951 Crick had made several clever theoretical contributions to crystallography. He had for example helped to show what kind of diffraction pattern a helical molecule would make. But his doctoral research lacked focus.

When Watson arrived at the Cavendish he was soon offered an office with Crick. Despite differences in age, nationality and personal experience they got on well and discovered a common interest in the structure of DNA. That they should have considered it an important problem is not at all surprising. That they believed an approach had to be made via crystallography is also not surprising. What is remarkable is that they felt they could tackle it themselves, along with their appointed tasks, without actually taking any more photographs. Partly this was a matter of temperament. They had the self-assurance to utilise the results of other people's work, believing not only that no-one would mind, but also that new significance could be extracted from them that others had missed. Partly too they picked a method which allowed them to make informed guesses, and which had already paid off for the American chemist, Linus Pauling, in his own work on proteins. The alternative would have involved extremely elaborate and time-consuming mathematical transformations of crystallographic data. This method was eventually used by Rosalind Franklin, working at King's College, London.

She had worked for a number of years in Paris. In 1951 she was recruited by Professor John Randall to join his biophysics group. Her understanding was that the crystallographic analysis of DNA was essentially her responsibility, because of her thorough grounding in the subject. Other researchers in the laboratory, amongst them Maurice Wilkins, who had been working on DNA there for several years, seem to have had much greater collaboration in mind. In the event personal relations between them were not a success and communication about research was minimal. Wilkins became increasingly irritated at what he believed to be Franklin's slow progress with the DNA analysis, pursuing rather unlikely possibilities, whilst other researchers, like Pauling in the United States, let their interest in DNA be known. In fact we now know that her disciplined consideration of all the interpretations was on the point of paying off. Had Watson and Crick not published their model in the spring of 1953, Rosalind Franklin would probably have done so, just a few weeks later.

In the autumn of 1951 she gave a seminar on her work to date, which James Watson attended. Back in Cambridge he and Crick tried to use what information Watson could remember to build a structural model. When the King's researchers arrived to inspect it they quickly showed that it was obviously wrong. Despite this embarrassing reverse Watson and Crick quietly maintained their interest in DNA, although they abandoned their model-making. In the summer of 1952 they met Erwin Chargaff on a visit to Cambridge. His detailed analytic work, using the newly developed technique of paper chromatography, had shown that, contrary to the tetranucleotide hypothesis, the four different bases in DNA are not always to be found in the same amounts. Chargaff's work also showed that the ratio of adenine to thymine is 1:1 and so is that of guanine to cytosine. The significance of this fact only emerged later, but it played a vital role.

In the autumn of 1952 the researchers in Cambridge learnt that Linus Pauling had actually started work on the DNA structure problem. In January 1953 he sent his son, Peter, then in the Cavendish, a manuscript containing his proposed solution. Quick inspection showed that it contained an elementary error, but Watson and Crick were re-energised to start building models again.

Watson began by trying to construct a two-chain model but with the chains of linked phosphates and deoxyribose sugars on the inside. Crick felt this could not be right, because it did not fit well with some of the chemical data. For example some of the work at King's College, London, and earlier studies elsewhere on the way in which DNA took up water, suggested that the bases had to be on the inside, and consequently the two, or possibly three, chains of linked phosphates and sugars on the outside. But if this was so, the problem then was to see how the bases, each shaped differently, could fit in a regular way at the centre of the molecule.

So the next attempt produced a model which paired like bases in the middle. This was quickly criticised by an American chemist,

Jerry Donohue, a former colleague of Pauling then working at the Cavendish, who pointed out that Watson was using the wrong chemical form of the bases. In nature their structure is slightly different. This advice was very helpful because it greatly reduced the number of possibilities and made it easier to see how the bases could be joined together. Now the question of how such bases could pair up had already been discussed by Crick with a young mathematician with an interest in biology, John Griffith. He had been asked to calculate what was the most likely combination. His answer was that adenine could join to thymine and guanine to cytosine with relatively weak hydrogen bonds. This did not fit with Watson's and Crick's preconceptions at the time and the advice remained unutilised until 1953.

Symmetry considerations suggested, so Crick argued, that the pairing of like bases was unlikely, so Watson next began to consider how to construct a model with unlike bases at its centre. One morning he noticed, using simple cardboard cut-outs, that if you placed adenine with thymine and guanine with cytosine then the resulting composite units each turned out to have the same shape. Here perhaps was the key to regular packing in the space between helical phosphate–sugar chains? Donohue confirmed that this was chemically plausible. After all Griffith's less expert comments had indicated this already. And Crick too could see no crystallographic objection. Moreover the exciting thing was that if the bases paired up in this way, then Chargaff's experimental result which showed constant ratios in the base composition of DNA would be explained. This tantalising finding, published in 1950, suddenly took on new meaning.

At this point in mid-February Crick took over the actual model-building, using more accurately made metal structures to represent the subunits. Everything had to fit together as exactly as possible, to be consistent with what was known of the chemistry of the sub-units and the bonds that linked them together and with the data in the X-ray diffraction patterns. By the end of the week of 28 February 1953 they had assembled a two-stranded model in which ten complementary pairs of bases, joined at the centre, were stacked one above the other in each complete turn of the helix. Moreover the complementary base-pairing (adenine always with thymine, guanine always with cytosine) meant that if the two strands split apart, by breaking the weak bonds between the base-pairs, then two exact copies of the original molecule could be built up in the cell by using each separated strand as a template. Thus not only did the double helix incorporate all the chemical

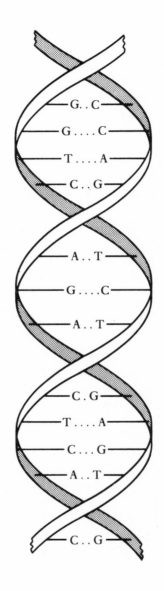

A simplified view of the structure of DNA proposed by Crick and Watson. The sugar-phosphate backbones that form the double helix are represented as ribbons. These are held together by the weak bonds between pairs of bases that span the central axis. Adenine (A) only pairs with thymine (T) and guanine (G) only pairs with cytosine (C).

and structural information, it also suggested how DNA could play its genetic role.

Other Cavendish colleagues were shown the model. Alexander Todd and his co-workers from the Chemistry Department came to inspect it. The King's College group were also informed and they too, in some surprise, said that it looked convincing. It is still not clear whether they were really aware how useful their data had been to Watson and Crick. Wilkins probably did know, Franklin probably did not. It was agreed with them that a paper by Watson and Crick should be sent off to the journal *Nature* for rapid publication, to be supported by papers from the London scientists.

After the first paper appeared in *Nature*, Watson and Crick drafted a second note, drawing attention specifically to the genetical implications of their model, lest others should rush these thoughts into print ahead of them. Watson had by this time written to Delbrück in the States and Pauling too had been put in the picture. Watson made a trip to Paris to discuss it with biologists at the Institut Pasteur. Shortly afterwards he returned to America. That summer he attended the annual meeting of the so-called phage group, the association of biologists using bacterial viruses for their genetic experiments. His presentation of the double helix was enthusiastically received.

But although scientists in the area we now call molecular biology saw its significance, many biologists took little immediate interest. Nor did it get the kind of media attention such a result would receive nowadays. It was described in the *News Chronicle* in the UK and in the *New York Times* but only briefly. Crick was keen to write about it in non-technical journals, Watson was much more hesitant, because for a surprisingly long while he feared it might turn out to be wrong.

The model stimulated two lines of scientific inquiry. One was essentially concerned with exploring the process of gene replication in yet more detail. This included a very clever experiment by two American scientists, Matthew Meselon and Franklin Stahl, in 1958, who found a way of labelling DNA strands with a radioactive isotope. They were able to show that with each generation the amount of DNA was reduced by half, which strongly supported the idea that DNA was copied by the separation of two complementary strands. This experiment was in effect a test of the double helix and it carried further the developing tradition within molecular biology of experiments, designed very precisely to test particular hypotheses. This had been going on in physics and chemistry for several centuries, but molecular biologists, some of them like Delbruck former physicists, thought of this as their modernising contribution to the life sciences.

The other line of inquiry concerned the genetic code. Living things are largely built of proteins of many different kinds. Genes, which specify how organisms are to be built, are made of DNA, which is chemically and structurally quite different. By the early 1950s, under the impact of the new fields of computing, cybernetics and information theory, it was common to speak of 'genetic information' and of a 'genetic code' which must relate the units of information in DNA to the sequential order of the building blocks of protein, the amino acids. One such code was proposed in 1952, but such exercises were necessarily highly speculative.

In this respect the double helix was very interesting, because clearly the coding elements had to be the bases. They were the only variable elements in the molecule; their sequence had presumably to be a coded message to the cell to make specific proteins. A great deal of theoretical effort was devoted in the 1950s to considering what kind of code was used. One of the key contributors to this discussion was Francis Crick. He was very influential in considering what kinds of code could be ruled out on theoretical grounds. He also proposed, just as a theoretical physicists might do, the existence of a special class of 'adaptor' molecules, which seemed to be necessary for the assembly of proteins as chains of amino acids. Subsequent experiments confirmed their existence. He also gave rise to the infamous phrase 'the central dogma', which held that genetic information could only flow from DNA to RNA (a related nucleic acid) to protein. Crick proposed this as a simplifying idea, that should not be called into question, in thinking about how proteins are made. The phrase is also an ironic reference to dogma in theology, which Crick as an atheist of long standing, found amusing. He also helped to develop an ingenious series of genetic experiments, using viruses, to determine how many bases made up any one unit of encoded genetic information. By 1960 it was clear that the coding units were groups of three bases, that they did not overlap and that there was no punctuation within the message.

Then in 1961 a biochemical system was devised that allowed each unit of the code to be deciphered. The race to decipher all sixty-four of them then began. Crick's Croonian Lecture to the Royal Society in 1966, which reviewed all the most recent results, can be taken as a triumphant closing speech on these years of work on the code. By this time the achievements of the 1950s, of which that by Watson and Crick is the best known, had ramified in many different directions and the field of molecular biology had achieved the predominance within the life sciences which it retains today.

In 1962 Watson, Crick and Wilkins shared a Nobel prize for their work on DNA. Their Cambridge colleagues who had been working on the analysis of protein, Max Perutz and John Kendrew, were similarly honoured in the same year. Rosalind Franklin had left King's College London in 1953 to join another crystallography group. By agreement, with Professor Randall she abandoned her interest in DNA, which Wilkins picked up. She turned instead to the analysis of tobacco mosaic virus. Her friendship with Crick and his family strengthened. However, her very promising career ended tragically in 1958 when she died of cancer. When the Nobel prize committee met in 1962 they were therefore unable to consider her for the award, as they could the living scientists.

BIOGRAPHIES

Chapter 1: Aristotle
Professor Geoffrey Lloyd was born in 1933 and educated at King's College, Cambridge, where he read classics and gained a PhD in 1958. Throughout his career he has maintained strong links with Cambridge University; he lectured in Classics from 1967–74, was Senior Tutor at King's from 1969–73 and was made Reader in Ancient Philosophy and Science in 1974. He travelled to the USA in 1981 as Visiting Professor at Stanford University and from 1983–84 at Berkeley. In 1983 he was appointed Professor of Ancient Philosophy and Science at Cambridge University. He is author of a number of books, including *Early Greek Science: Thales to Aristotle* (Chatto, 1970) and *Science and Morality in Greco-Roman Antiquity* (Cambridge University Press, 1985). (See also booklist.)

Chapter 2: Ptolemy
John North, born in Cheltenham in 1934, is Professor of the History of Philosophy and the Exact Sciences at the University of Groningen, The Netherlands. Before moving there in 1977 he was at the University of Oxford, first as an undergraduate at Merton College, later as Nuffield Research Fellow in History and Philosophy of Science, and subsequently at the Museum of the History of Science. He has held the post of Visiting Professor in, among others, the Universities of Frankfurt, Aarhus, Minnesota and Austin. His books include *The Measure of the Universe* (Oxford University Press, 1965), *Horoscopes and History* (The Warburg Institute of the University of London, 1986) and *Chaucer's Universe* (Oxford University Press, 1987).

Chapter 3: Galileo
Born in London in 1920, *Colin Ronan* studied at Imperial College and University College, London, where he gained a degree in Astronomy and a Masters in History and Philosophy of Science. In 1949 he became a senior member of the Secretariat of the Royal Society and in 1960 began working as a freelance author and

editor. From 1965–85 he was editor of the *Journal of the British Astronomical Association* and since then has been project coordinator of the Science and Civilisation in China project at the Needham Research Institute in Cambridge. He is author of over 30 books including *Isaac Newton* (International Textbook Co., 1969) and *The Cambridge Illustrated History of the World's Science* (Cambridge University Press, 1983). (See also booklist.)

Chapter 4: Kepler

Dr Jim Bennett was born in 1947 in Belfast. He was educated at the University of Cambridge where he obtained both a degree and a PhD in History of Science. His former posts include Lecturer in History and Philosophy of Science at the University of Aberdeen and Curator of Astronomy at the National Maritime Museum, Greenwich. At present he is a Fellow of Churchill College and Curator of the Whipple Museum of the History of Science at the University of Cambridge. He is author of *The Mathematical Science of Christopher Wren* (Cambridge University Press, 1982).

Chapter 5: Harvey

Dr Andrew Cunningham, born in London in 1945, studied at the University of Oxford where he received his first degree in History. In 1974 he gained a PhD in History of Medicine at the University of London. He is currently Wellcome Lecturer in the History of Medicine at the University of Cambridge.

Chapter 6: Newton

Professor Alfred Rupert Hall was born at Stoke-on-Trent in 1920 and educated at Christ's College, Cambridge. He took both his first degree and PhD in History. After lecturing in History of Science until 1959, he travelled to UCLA where he was first a medical research historian and then Professor of Philosophy. In 1961 he became Professor of History and Logic of Science at Indiana University and in 1963 returned to Britain to take up the chair in History of Science and Technology at Imperial College, London. His books include *Philosophers at War* (Cambridge University Press, 1980) and *Physic and Philanthropy: A History of the Wellcome Trust* (Cambridge University Press, 1986). (See also booklist).

Chapter 7: Priestley

Born in Leeds in 1946, *John Christie* graduated in History at the University of Edinburgh where he remained for 3 years to research into the history of Scottish science. In 1973 he moved to the University of Leeds, first as Research fellow, then as Lecturer in History and Philosophy of Science. He has co-edited two books: *Martyr of Science* (HMSO – Royal Scottish Museum, 1984) and *The Figural and the Literal* (Manchester University Press, 1987).

Chapter 8: Lavoisier

Professor Maurice Crosland was born in London in 1931 and studied at the University of London where, in 1959, he gained a PhD in History of Science. From 1963 he lectured in History of Science at the University of Leeds. He was Visiting Professor at the University of California, Berkeley, in 1967, at Cornell University from 1967–68, and at the University of Pennsylvania in 1971. In 1974 he was made Professor of the History of Science and Director of the Unit for the History, Philosophy and Social Relations of Science at the University of Kent. He is author of *The Society of Arcueil: a View of French Science at the Time of Napoleon I* (Heinemann, 1967) and *Gay-Lussac, scientist and bourgeois* (Cambridge U.P., 1978). (See also booklist.)

Chapter 9: Watt

Professor Donald Cardwell was born in Gibraltar in 1919, studied physics at the University of London and gained his PhD in 1949. He lectured at Leeds University from 1960–1963, before moving to The University of Manchester Institute of Science and Technology to become first Reader and then, in 1974, Professor of Science and Technology. Since 1984 he has been Emeritus Professor. He is author of *Steam Power in the Eighteenth Century* (Sheed and Ward, 1963) and *From Watt to Clausius: the Rise of Thermodynamics in the Early Industrial Age* (Heinemann, 1971).

Chapter 10: Faraday

Born in 1936, *Dr David Knight* went to Keble College, Oxford, to read Chemistry in 1957. He then began research on the history of chemistry and obtained a D.Phil. in 1964. In the same year he moved to Durham University where he has remained, teaching History of Science. He has been editor of *The British Journal for the History of Science* since 1981. His most recent book is *The Age of Science* (Basil Blackwell, 1986).

Chapter 11: Darwin

Dr John Durant was born in Norwich in 1950. He took his first degree in Zoology and his second in the History of Science at Queens' College, Cambridge. Since then, he has taught mature students, first at the University College, Swansea and, more recently, in the Department for External Studies at the University of Oxford, where he is currently Staff Tutor in Biological Sciences. He edited and co-authored *Darwinism and Divinity: Essays on Evolution and Religious Belief* (Blackwell, 1985).

Chapter 12: Pasteur

Dr William Bynum was born in Abilene, Texas, in 1943. After qualifying as a medical doctor at Yale University he came to Britain where he gained a PhD in History of Science from the University of Cambridge. He has lived in Britain since 1970 and is head of the Academic Unit at the Wellcome Institute for the History of Medicine and Reader in History of Medicine at University College, London. He is co-editor of and contributor to *William Hunter and the Eighteenth Century Medical World* (Cambridge University Press, 1985), *The Anatomy of Madness* (Tavistock Publications, 1985) and *Medical Fringe and Medical Orthodoxy* (Croom Helm, 1986).

Chapter 13: Einstein

Professor Clive Kilmister was born in Epping, Essex, in 1924 and read Mathematics at Queen Mary College, London. He received his PhD in 1950 and moved to King's College London where he became first Lecturer, then Reader and finally Professor of Mathematics in 1966. He retired from this post in 1984 but has remained Gresham Professor of Geometry. His books include *The environment in modern physics: a study in relativistic mechanics* (English U.P., 1965) and *The nature of the Universe* (Thames & Hudson, 1971). (See also booklist.)

Chapter 14: Bohr

Dr John Hendry, born in 1952, studied Mathematics at the University of Cambridge before moving to Imperial College, London, where he obtained an MSc and PhD in History of Science. He is currently Fellow in Business History at the London Business School. He is author of *Cambridge Physics in the Thirties* (Hilger, 1984) and *James Clerk Maxwell and the Theory of the Electromagnetic Field* (Hilger, 1986). (See also booklist.)

Chapter 15: Turing

Dr Andrew Hodges was born in London in 1949 and took his first degree in Mathematics at the University of Cambridge. His post-graduate and postdoctoral research have been in twistor theory – a new mathematical approach that gives a radically new way of looking at space, time and matter. He is presently attached to the Mathematical Institute, Oxford University. He is also biographer of Alan Turing. (See book list.)

Chapter 16: Watson and Crick

Born in 1950 at Bedford, *Dr Edward Yoxen* was a student at King's College, Cambridge where he studied Engineering and subsequently obtained a PhD in History and Philosophy of Science. In 1976 he was appointed Lecturer in Science and Technology Policy, University of Manchester, where he has remained since. He is author of *The Gene Business* (Pan, 1983) and *Unnatural Selection? Coming to terms with the new genetics* (Heinemann, 1986).

Roy Porter: Introduction

Dr Roy Porter was born in 1946 and studied at Cambridge. He became a Fellow of Christ's College and then of Churchill College, Cambridge, and University Lecturer in History. In 1979, he moved to the Wellcome Institute for the History of Medicine in London, where he is now Senior Lecturer in the Social History of Medicine. He has co-edited the *Dictionary of the History of Science* (Macmillan, 1981) and *The Anatomy of Madness* (2 vols, Tavistock, 1985) and is also the author of *English Society in the Eighteenth Century* (Penguin, 1982).

BOOK LIST

General

For a good introduction and chronological outline:

RONAN, C. *The Cambridge illustrated history of the world's science* Cambridge U.P., 1983.
Two good introductions to recent books on the subject are:
CORSI, P. and WEINDLING, P. eds *Information sources in the history of science and medicine* Butterworth, 1982.
DURBIN, P.T. ed. Collier MacMillan, 1980; n.e. pbk 1985.
An annual bibliography of new works in the history of science and medicine is available in *ISIS*, a journal of the history of science, published in the United States.

Chapter 1: Aristotle

Works by Aristotle

Aristotle, the basic works ed. R. McKeon. New York: Random, 1941. o.p.
Complete works of Aristotle (Revised Oxford transl.) ed. J. Barnes. Princeton: Princeton U.P., 1984.

Works about Aristotle

ACKRIL, J. L. *Aristotle the philosopher* Oxford U.P., 1981.
ALLEN, D. J. *The philosophy of Aristotle* Oxford U.P., 2nd rev. edn, pbk, 1970.
BARNES, J. *Aristotle* (Past Masters Series) Oxford U.P., 1982.
LLOYD, G. E. R. *Aristotle, the growth and structure of his thought* Cambridge U.P., 1968.
ROSS, W. D. *Aristotle* Methuen, 5th edn repr., 1968.

Chapter 2: Ptolemy

Works by Ptolemy

The Almagest trans. G. J. Toomer. Duckworth, 1984.
L'Optique de Claude Ptolémée ed. A. Lejuene. Louvain: U. Louvain Pr., 1956. o.p.
Tetrabiblos ed. F. E. Robbins (Loeb Classical Library No. 435) Cambridge, Mass: Harvard U.P., 1980.

Works about Ptolemy

PEDERSEN, O. *A survey of the Almagest* Odense: Odense U.P., 1974. o.p.

Chapter 3: Galileo

Works by Galileo

Dialogue concerning two new sciences transl. H. Crew and A. de Salvio. New York & London: Dover, 1914, repr. 1954.
Dialogue concerning the two chief world systems – Ptolemaic and Copernican transl. S. Drake. Berkeley and London: U. California Press, 2nd. rev. edn, 1967.
GALILEI, G., GRASSI, H., GUIDUCCI, M. and KEPLER, J. *The controversy on the comets of 1618* transl. S. Drake and C. D. O'Malley. Oxford U.P., 1961, o.p.

Works about Galileo

DRAKE, S. *Galileo at work: his scientific biography* Chicago & London: U. Chicago Pr., new edn, pbk, 1981.
KAPLON, M. F. ed. *Homage to Galileo* Cambridge, Mass.: MIT Pr., 1965.
RONAN, C. A. *Galileo* Weidenfeld, 1974, o.p.
SHEA, W. R. *Galileo's intellectual revolution* Macmillan, 1972. o.p.

Chapter 4: Kepler

Works by Kepler

Mysterium cosmographicum: the secret of the universe transl. A. M. Duncan. New York: Abaris, 1981.
Great books of the western world Vol. 16: *Harmonies of the world* Chicago and London: Encyclopedia Britannica, 1952.

Works about Kepler

BEER, A. and P. eds *Vistas in astronomy* Vol 18: *Kepler, four hundred years* (Proceedings of a conference held in honour of Johannes Kepler) Pergamon, 1975.

CASPAR, M. *Kepler* transl. C. D. Hellman. New York: Abelard-Schuman, 1959. o.p.

JARDINE, N. *The birth of history and philosophy of science* Cambridge U.P., 1984.

KOYRÉ, A. *The astronomical revolution: Copernicus, Kepler, Borelli* transl. R. E. W. Maddison. Methuen, new edn, 1980.

Chapter 5: Harvey

Works by Harvey

The circulation of the blood transl. and ed. K. J. Franklin. (Everyman) Dent.

Works about Harvey

BYLEBYL, J. *William Harvey and his age* Johns Hopkin U.P., 1979. o.p.

KEYNES, C. *The life of William Harvey*, Oxford U.P., 1966.

Chapter 6: Newton

Works by Newton

Optics: or, A treatise of the reflections, refractions, inflections and colours of light New York and London: Dover, 1952.

Works about Newton

HALL, A. R. *The revolution in science, 1500–1750* Longman, 1983.

WESTFALL, R. S. *Never at rest: a biography of Isaac Newton* Cambridge U.P., new edn, pbk, 1983.

Chapter 7: Priestley

Works by Priestley

The autobiography of Joseph Priestley. Memoirs written by himself, an account of further discoveries in air Adams & Dart, 1970. o.p.

Works about Priestley

GIBBS, F. W. *Joseph Priestley: adventurer in science and champion of truth* Nelson, 1965. o.p.
MCLACHLAN, J. *Joseph Priestley, man of science, 1733–1804: an iconography of a great Yorkshireman* Merlin, 1983 o.p.
ORANGE, A. D. *Joseph Priestley: an illustrated life* Shire, 1974.
THORNE, T. E. *Joseph Priestley* New York: AMS Pr., 1981.

Chapter 8: Lavoisier

Works by Lavoisier

Elements of chemistry transl. R. Kerr. New York and London: Dover, new edn, pbk, 1984.

Works about Lavoisier

Dictionary of scientific biography (vol. 8) ed. C. C. Gillispie. (*Lavoisier* by H. Guerlac) New York: Scribners, new edn 1985.
CROSLAND, M. P. *Historical studies in the language of chemistry* New York and London: Dover, pbk, 1980.
GILLISPIE, C. C. *The edge of objectivity* (Chapter 6: The rationalisation of matter) Princeton: Princeton U.P.; Oxford U.P., 1960. o.p.
HOLMES, F. L. *Lavoisier and the chemistry of life: an exploration of scientific creativity* Madison, Wisconsin: U. Wisconsin Pr., 1984.

Chapter 9: Watt

Works by Watt

BOULTON, M. and WATT, J. *The selected papers of Boulton and Watt Vol. 1: the engine partnership, 1775–1825* ed. J. Tann. Cambridge, Mass.: MIT Pr., 1981.
WATT, J. and BLACK, J. *Partners in science: the letters of James Watt and Joseph Black* ed. E. Robinson and D. McKie. Cambridge, Mass.: Harvard U.P., 1969: London: Constable, 1970 o.p.

Works about Watt

CARDWELL, D. S. L. *Technology, science and history* Heinemann, 1972. o.p.
DICKINSON, H. W. *James Watt, craftsman and engineer* New York: Kelley, new edn of 1936 edn, 1978.
DICKINSON, H. W. *A short history of the steam engine* F. Cass, 2nd edn, 1963.
HILLS, R. L. *Power in the industrial revolution* Manchester U.P., 1970. o.p.

Chapter 10: Faraday

Works by Faraday

Chemical Manipulation (1827) London, 1827.
Experimental researches in electricity (3 vols) Repr. from *Philosophical Transactions*, 1839–55.
Experimental researches in chemistry and physics Repr. from *Philosophical Transactions*, 1859.
A course of six lectures on the various forces of matter and *A course of six lectures on the chemical history of a candle* ed. W. Crookes, 1860.
Faraday's Diary (7 vols) ed. T. Martin. Royal Inst., 1932–36.
WILLIAMS, L. P. *et al.* eds *The selected correspondence of Michael Faraday* (2 vols) Cambridge U.P., 1971. o.p.

Works about Faraday

WILLIAMS, L. P. *Michael Faraday: a biography* Chapman & Hall, 1965. o.p.
GOODING, D. and JAMES, F. A. J. L. eds *Faraday rediscovered* Macmillan, 1985.
RUSSELL, C. *Science and social change, 1700–1900* Macmillan, 1983.

Chapter 11: Darwin

Works by Darwin

On the origin of species by means of natural selection, or the preservation of favoured races in the struggle for life ed. J. W. Burrow. Penguin, 1982.
The voyage of the 'Beagle' (Everyman) Dent, 1983.
The expression of the emotions in man and animals University of Chicago Press, 1965.

Works about Darwin

BOWLER, P. *Evolution, the history of an idea* Berkeley, Los Angeles and London: U. California Pr., 1984.
GRUBER, H. E. *Darwin on man: a psychological study of scientific creativity* Chicago: U. Chicago Pr., 2nd edn, pbk 1981.
MILLER, J. and VAN LOON, B. *Darwin for beginners* Writers and Readers Pub. Co-op.; Unwin, new edn, pbk, 1986.
YOUNG, R. M. *Darwin's metaphor: nature's place in Victorian culture* Cambridge U.P., 1985.

Chapter 12: Pasteur

Works by Pasteur

Oeuvres de Pasteur (7 vols) ed. P. Vallery-Radot. Paris, 1922–1939, o.p.
Studies on fermentation transl. F. Faulkner and D. C. Robb. Macmillan, 1879. o.p.

Works about Pasteur

DUCLAUX, E. *Pasteur: the history of a mind* transl. E. F. Smith and F. Hedges. Scarecrow Pr., new edn, 1974.
DUBOS, R. *Louis Pasteur: free lance of science* Da Capo Pr., 1986.
VALLERY-RADOT, R. *The life of Pasteur* transl. Mrs R. L. Devonshire. Constable, 1901.
Dictionary of scientific biography ed. C. C. Gillispie (Louis Pasteur by G. Geison) New York: Scribners, new edn, 1985, vol. 10.

Chapter 13: Einstein

Works by Einstein

The meaning of relativity Chapman and Hall, 1967.
Out of my later years Greenwood, 1950.
Philosophical problems of twentieth century physics Central Books, 1984.
Sidelights on relativity Dover, 1983.
EINSTEIN, A. and INFELD, L. *Evolution of physics: the growth of ideas from early concepts to relativity and quanta* Cambridge U.P., 1971.

Works about Einstein

KILMISTER, C. W. *General theory of relativity* Pergamon, 1970.
KILMISTER, C. W. *Special theory of relativity* Pergamon, 1973.
STEPHENSON, G and KILMISTER, C. W. *Special relativity for physicists* Longmans, 1958. o.p.

Chapter 14: Bohr

Works by Bohr

Atomic theory and the description of nature Cambridge U.P., 1961. o.p.
Atomic physics and human knowledge New York: Wiley; London: Chapman & Hall, 1958. o.p.
Essays, 1956–1962, on atomic physics and human knowledge New York & London: Interscience. o.p.

Works about Bohr

FOLSE, H. *The philosophy of Niels Bohr: the framework of complementarity* Amsterdam: North-Holland, 1985.
FRENCH, A. P. and KENNEDY, P. J. eds *Niels Bohr: a centenary volume* Cambridge, Mass. and London: Harvard U.P., 1985.
HENDRY, J. *The creation of quantum mechanics and the Bohr-Pauli dialogue* Dordrecht, Boston & Lancaster: D. Reidel, 1984.
ROZENTAL, D. ed. *Niels Bohr: his life and work as seen by his friends and colleagues* Amsterdam: North-Holland, 1967.

Chapter 15: Turing

Works by Turing

The automatic computing engine: papers by A. Turing and M. Woodger (Charles Babbage Inst. repr. for History of Computing) Cambridge, Mass.: MIT Pr., 1985.

Works about Turing

HODGES, A. *The enigma of intelligence* Counterpoint, 1985.
HOFSTADTER, D. R. *Gödel, Escher, Bach: an eternal golden braid* Harvester Pr., 1979; Penguin, 1980.
HOFSTADTER, D. R. and DENNETT, D. C. *The mind's I: fantasies and reflections on self and soul* Harvester Pr., 1981; Penguin, 1982.
STEEN, L. A. *Mathematics today* New York: Springer, 1978, o.p.; Random, pbk, 1980.

Chapter 16: Watson and Crick

Works by Watson

The double helix: A personal account of the discovery of the structure of DNA Weidenfeld, 1981.
The molecular biology of the gene Benjamin/Cummings, 1976.
WATSON, J. and TOOZE, J. eds *The DNA story: a documentary history of gene cloning* W. H. Freeman, 1981; new edn, pbk, 1983.

Works by Crick

Of molecules and men Seattle: U. Washington Pr., 1966.
The double helix: a personal view *Nature* **248** 766–769, 1974.
The genetic code *Proceedings of the Royal Society, B* **167**, 331–347, 1967.

Works about Watson and Crick

JUDSON, H. F. *The eighth day of creation: the makers of the revolution in biology* Cape, 1979.

OLBY, R. C. *The path to the double helix* Macmillan, 1974. o.p.

PORTUGAL, F. H. and COHEN, J. S. *A century of DNA: a history of the discovery of the structure and function of the genetic substance* Cambridge Mass.: MIT Pr., 1977.

SAYRE, A. *Rosalind Franklin and DNA* New York: Norton, 1975.

Acknowledgements

Acknowledgement is due to the following, whose permission is required for multiple reproduction:

SECKER (MARTIN) & WARBURG LTD for the extract from *Aubrey's brief lives* ed. Clive Lawson Dick (Secker (Martin) & Warburg Ltd).

Picture Credits

American Institute of Physics, Nils Bohr Library (Einstein); A.C. Barrington Brown (from *The Double Helix* by J.D. Watson, pub. Atheneum, New York); BBC Hulton Picture Library (Aristotle, Watt, Faraday, Pasteur); Biblioteca Marucelliania, Florence (Galileo); Photographie Bulloz, Paris (Harvey); Camera Press (Watson & Crick); Deutsches Museum, Munich (Kepler); National Portrait Gallery (Lavoisier, Newton, Priestley, Darwin); Ann Ronan Picture Library (Turing); Royal Society (Ptolemy).

The cover picture shows a detail from the painting *Experiment with the Air Pump* by Joseph Wright of Derby (National Gallery).

Index